D0839989

ELECTRIC MOTOR
PROFESSIONAL REFERENCE

Paul Rosenberg

Created exclusively
for DeWALT by:

PAL
publications®

FEB 06 ⌶

www.palpublications.com
1-800-246-2175

Titles Available From DeWALT

DeWALT Trade Reference Series

Construction Professional Reference

Datacom Professional Reference

Electric Motor Professional Reference

Electrical Estimating Professional Reference

Electrical Professional Reference

HVAC Professional Reference

Lighting & Maintenance Professional Reference

Plumbing Professional Reference

Referencia profesional sobre la industria eléctrica

Security, Sound & Video Professional Reference

Wiring Diagrams Professional Reference

DeWALT Exam and Certification Series

Electrical Licensing Exam Guide

HVAC Technician Certification Exam Guide

This Book Belongs To:

Name:_____

Company: _____

Title: _____

Department: _____

Company Address: _____

_____

_____

Company Phone: _____

Home Phone: _____

Pal Publications, Inc.
374 Circle of Progress
Pottstown, PA 19464-3810

ISBN 0-9759709-6-8

09 08 07 06 05   5 4 3 2 1

Printed in the United States of America

A Note To Our Customers

We have manufactured this book to the highest quality standards possible. The cover is made of a flexible, durable and water-resistant material able to withstand the toughest on-the-job conditions. We also utilize the Otabind process which allows this book to lay flatter than traditional paperback books that tend to snap shut while in use.

Electric Motor Professional Reference is not a substitute for the National Electric Code®. National Electric Code® and NEC® are registered trademarks of the National Fire Protection Association, Inc. Quincy, MA.

Preface

Motor work was the thing I excelled at the most in trade school, yet I have never written specifically on the subject. So, putting together this book for DEWALT was a pleasure for me. It was also gratifying to see that the book developed as we had hoped — a small book with almost everything an electrician really needs to know about electric motors.

Obviously, any attempt to cover electric motors in a single book requires that certain material is excluded. In this book, I chose to cover the material that is used most by people who install and maintain electric motors. This book does not cover the internal construction of electric motors, except to the extent necessary to address internal wiring or troubleshooting. But it does cover all of the important aspects of this work: designing, installing, maintenance, power transmission, troubleshooting, connections and more. These are the things that matter most in motor work, and I am confident that we've covered the central technologies well.

Naturally, there may be some aspects of motor work that I have overlooked, or that are not covered in sufficient depth for some readers. I continuously monitor the industry and will update this book on a regular basis. I will also attempt to include additional material suggested by readers and to keep pace with developments in the electrical trade.

Best wishes,
Paul Rosenberg

Chapter 1 –
Formulas and Fundamentals. **1-1**

Inductance 1-2

Capacitance 1-4

Mechanical and Electrical Degrees 1-5

Poles 1-6

Synchronous Speed 1-7

Slip 1-7

Speed Regulation 1-7

AC Motor Torque 1-8

Wound-Rotor Motor Torque 1-9

DC Motor Torque 1-10

Squirrel-Cage Motors 1-11

Wound-Rotor Motors 1-12

Synchronous Motors 1-13

DC Motors 1-14

Ohm's Law for Alternating Current 1-15

Ohm's Law for Direct Current 1-16

Power Factor 1-16

Single-Phase Power 1-16

Three-Phase AC Circuits and
  the Utilization of Power 1-17

Wye Connection 1-17

Delta Connection 1-17

Four-Wire System 1-17

Types of Power 1-18

True Power and Apparent Power 1-19

Power Factor Formula 1-19

AC/DC Power Formulas. 1-20

Horsepower Formulas 1-21

Efficiency Formulas 1-21

Voltage Unbalance 1-21

Temperature Conversions 1-21

Voltage Drop Formulas 1-22

Conductor Length/Voltage Drop 1-22

Conductor Size/Voltage Drop 1-22

Formulas for Sine Waves. 1-23

Calculating Root-Mean-Square (RMS) 1-23

Summary of Series, Parallel and
  Combination Circuits 1-24

Chapter 2 – *Motor Construction* 2-1

Capacitor-Start Motor 2-1

Operation of Capacitor-Start Motor 2-2

Four-Pole Capacitor-Start Motor. 2-3

Four-Pole, Two-Circuit Capacitor-Start Motor 2-4

Six-Pole Capacitor-Start Motor. 2-5

Dual-Voltage Motors 2-6

Single-Voltage Capacitor-Start Motor 2-8

Dual-Voltage Capacitor-Start Motor 2-9

Reversing Single-Phase Motors 2-11

Jumpers 2-12

Six-Pole Motor 2-15

Reversing Split-Phase Motors. 2-16

Permanent-Split Capacitor Motor 2-17

Two-Value Capacitor Motor 2-18

Synchronous Motors 2-19

Shaded-Pole Motors 2-22

Capacitor Motors 2-27

Multi-Speed Single-Phase Motors 2-29

Universal Motors 2-34

Connections for Multi-Speed Squirrel-Cage Motors . . . 2-39

Wye Three-Phase Motors 2-44

    Reversing Direction 2-45

    Standard Markings 2-46

    Series-Star 2-48

    Four-Pole 2-50

Wye-Wound Motor for Use on 240/480 Volts 2-52

Numbering of Dual-Voltage Wye-Wound Motor 2-52

Delta Three-Phase Motors 2-53

    Standard Markings 2-53

    Four-Pole Series Delta 2-55

DC Motors 2-56

    Two-Pole Compound-Interpole 2-56

    Brush Positions 2-58

    Two-Pole Shunt 2-61

    Two-Pole Series 2-63

    Compound-Interpole 2-65

    Cumulative Compound 2-69

Chapter 3 –
Connections and Diagrams 3-1

Delta-Wound Motor for Use on 240/480 Volts 3-1

Wye-Wound Motor for Use on 240/480 Volts. 3-1

Split-Phase Motors. 3-2

Split-Phase Motor Rotation 3-3

Single-Voltage, 3φ, Wye-Connected Motor 3-4

Single-Voltage, 3φ, Delta-Connected Motor 3-4

Dual-Voltage, 3φ, Wye-Connected Motors 3-5

Two-Phase, AC Motors 3-6

Connections for A Two-Speed,
  Constant Horsepower, One Winding Motor 3-7

Connections for A Two-Speed,
  Constant Torque, One Winding Motor. 3-8

Connections for A Two-Speed,
  Variable Torque, One Winding Motor. 3-9

Capacitor-Start-Capacitor-Run Motor. 3-10

Reversing A Capacitor-Start Motor 3-10

Capacitor-Start Motor Voltage Connections. 3-11

Two-Speed Capacitor Motor 3-11

Wound-Rotor Motor Schematic 3-12

Reversing Split-Phase Motors. 3-12

Typical Motor Starter Diagram. 3-13

Step-Down Transformer Motor Control 3-13

Motor Control Circuits 3-14

  Stop-Start Stations, Jogging, Plugging. 3-15

  Low-Voltage Control, Magnetic Starters 3-17

Magnetic Starters, Starting Compensator 3-19

Reversing Type Starter......................... 3-21

Two-Speed Motor with Push-Button.............. 3-23

Manual Type Starting Compensator 3-25

Resistance and Faceplate Starters 3-27

Regulating Rheostat........................... 3-29

Adjustable-Speed Capacitor Motors.............. 3-31

Multi-Speed Motor Control 3-33

Synchronous Motors, Exciter-Field Rheostat 3-38

Reduced-Voltage Starting Systems 3-44

Changing Speeds.............................. 3-47

Starting, Stopping, Jogging..................... 3-53

DC Series-Wound Motor 3-54

DC Shunt-Wound Motor 3-54

DC Compound-Wound Motor...................... 3-54

DC Motor Connections 3-55

Series-Wound................................. 3-55

Shunt-Wound, Compound-Wound 3-57

Compound Generators 3-61

Commutating-Pole............................ 3-63

DC Motor Control Circuits 3-64

Variable Resistance Speed Control 3-65

Speed-Regulating Faceplate Starter.............. 3-69

Speed-Regulating Rheostat Control 3-71

Step-by-Step Resistance Regulation 3-73

Constant-Speed Shunt and Compound-Wound
Reversible Motor Control. 3-75

Chapter 4 – *Design and Installation* . . . 4-1

Designing Motor Circuits 4-2

Tips on Selecting Motors. 4-3

Facts to Consider. 4-3

Summary of Motor Applications 4-4

Three-Phase Motor Requirements 4-8

Direct Current Motor Requirements 4-9

1φ 115 V Motors and Circuits – 120 V System. 4-10

1φ 230 V Motors and Circuits – 240 V System. 4-11

3φ 230 V Motors and Circuits – 240 V System. 4-13

3φ 460 V Motors and Circuits – 480 V System. 4-16

DC Motors and Circuits 4-19

Control Ratings. 4-22

Standard Motor Sizes 4-25

Starting Methods: Squirrel-Cage Induction Motors 4-26

Setting Branch Circuit Protective Devices 4-27

Maximum OCPD 4-28

Standard Sizes of Fuses and CB's 4-28

Motor Power Formulas — Cost Savings 4-29

Heater Selections. 4-30

Heater Trip Characteristics 4-33

Heater Ambient Temperature Correction 4-34

Heating Element Specifications. 4-35

Heater Construction. 4-36

Motor Bearings and End Plates 4-37

Motor Disconnects. 4-38

Motor Feeder Tap Sizing 4-40

General Controller Fault Current Limits. 4-41

Horsepower Limits of Individual Motors. 4-41

Common Service Factors 4-42

Small Motor Guide. 4-43

DC Motor Performance Characteristics 4-44

Maximum Acceleration Time. 4-44

DC Motor Efficiencies, Continuous Rated,
   40°C Rise 4-45

Power Factor Improvement. 4-46

Typical Motor Power Factors. 4-47

Efficiency Formulas 4-47

Voltage Unbalance Formula. 4-47

Motor Torque (Inch Pounds-Force) 4-48

Horsepower to Torque Conversion 4-51

Locked Rotor Current 4-52

Nema Ratings of 60Hz AC Contactors. 4-53

Full-Load Currents: DC Motors. 4-54

Full-Load Currents: 1ɸ, AC Motors 4-54

Full-Load Currents: 3ɸ, AC Induction Motors. 4-55

General Effect of Voltage Variation on
   Direct Currect Motor Characteristics. 4-56

General Effect of Voltage and Frequency
   Variation on Induction Motor Characteristics 4-57

Voltage Variation Characteristics. 4-58

Frequency Variation Characteristics 4-58

Typical Motor Efficiencies 4-59

Types of Enclosures. 4-60

Enclosure Insulation, Temperature 4-60

Enclosure Ratings 4-61

Hazardous Locations 4-61

Enclosure Types 4-62

Frontal View of Typical Motor 4-64

Side View of Typical Motor 4-64

NEMA Motor Frame Dimensions. 4-65

Shaft Coupling Selections 4-66

Motor Frame Letters 4-67

Motor Frame Dimensions 4-68

Motor Frame Table. 4-70

Motor and Ampacity Ratings. 4-72

Locations 4-72

Grounding. 4-72

Overload Protection 4-72

Motor Circuit Conductors 4-75

Motor Controllers 4-76

Motor Control Circuits 4-77

Conductors Supplying Motors and Other Loads 4-78

Conductors Supplying Several Motors. 4-79

Short-Circuit and Ground-Fault Protection. 4-79

Adjustable-Speed Drives 4-80

Chapter 5 – *Maintenance* **5-1**

Motor Maintenance Operations. 5-2

Motor Repair and Service Record. 5-4

Semiannual Motor Maintenance Checklist 5-5

Annual Motor Maintenance Checklist 5-6

Locating Circuits. 5-8

Checking Capacitors 5-9

Unmarked 3ϕ Induction Motors. 5-10

Wye Or Delta Connection 5-10

Wye-Connected Motor 5-11

Delta-Connected Motor. 5-11

DC Motor Performance Characteristics 5-12

Maximum Acceleration Time. 5-12

AC Voltage Variation Characteristics. 5-13

AC Frequency Variation Characteristics 5-13

Phase Unbalance and Temperature Rise 5-14

Phase Unbalance Derating Factor 5-15

Single-Phasing Condition 5-16

Improper Phase Sequence (Phase Reversal) 5-17

Voltage Surge 5-17

Voltage Problems 5-18

Voltage Variance. 5-19

Acceptable AC Load Voltage Ranges (60Hz) 5-19

Voltage Unbalance. 5-20

Finding Voltage Unbalance 5-21

Motor Overcycling 5-22

Improper Ventilation. 5-22

Excessive Heat. 5-23

Motor Overloads. 5-24

Megohmmeter Connections 5-25

Ohmmeter Connections. 5-26

Insulation Spot Testing. 5-27

Dielectric Absorption Testing. 5-28

Polarization Index Values. 5-29

Insulation Step Voltage Testing 5-30

Chapter 6 – *Troubleshooting* 6-1

Contactor and Motor Starter Troubleshooting Guide. . . . 6-2

Direct Current Motor Troubleshooting Guide 6-4

Shaded Pole Motor Troubleshooting Guide 6-6

Split-Phase Motor Troubleshooting Guide 6-7

Three-Phase Motor Troubleshooting Guide 6-10

Faulty Solenoid Problems 6-12

Mechanical Start 6-14

Split-Phase Motor 6-14

Motor Current. 6-15

Defective Pole 6-16

Start Winding 6-17

Stator 6-18

Grounded Winding. 6-19

Disassembly 6-19

Shorted Coil 6-20

Opens 6-22

Interpole Polarity 6-24

Brushes. 6-25

Compound Motors. 6-26

DC Opens. 6-30

DC Grounds 6-33

Two-Pole Motors 6-36
Polarity 6-38
AC Grounds 6-41
AC Opens 6-45
Shorts 6-52
Fuses 6-53
Mechanical 6-55

Chapter 7 – *Power Transmission* 7-1

Method of Aligning Pulleys 7-1
Calculating Pulley Diameter....................... 7-2
Pulley and Gear Calculations...................... 7-3
Formulas for Finding Pulley Sizes 7-4
Formulas for Finding Gear Sizes 7-4
Formula to Determine Shaft Diameter.............. 7-5
Formula to Determine Belt Length 7-5
Gear Reducer Formulas........................... 7-6
Motor Torque Formulas 7-7
Adjusting Belt Tension 7-8
V-Belts... 7-9
V-Belt/Motor Size 7-9
Standard "V" Belt Lengths........................ 7-10
Horsepower Capacities of Light 4-Ply
    Nylon-Stitched Belts 7-12
Horsepower Capacities of Medium 4-Ply
    Nylon-Stitched Belts 7-12
Horsepower Capacities of
    Heavy 4-Ply Nylon-Stitched Belts 7-13

Horsepower Capacities of Medium
   4-Ply Woven Endless Cotton Belts 7-13

Horsepower Capacities Per Inch of Width
   of Regular Single-Ply Dacron Belts 7-14

Horsepower Capacities Per Inch of Width
   of Medium Single-Ply Dacron Belts 7-15

Horsepower Capacities Per Inch of Width
   of Light Single-Ply Dacron Belts 7-16

Horsepower Capacities of $\frac{3}{8}$" Diameter Wound
   Endless Round Belts 7-17

Correction Factors for Small Pulley Angles of
   Contact Less Than 180° 7-18

Correction Factors for Belt Variations 7-19

Horsepower Capacities of $\frac{1}{4}$"
   Diameter Braided Endless Round Belts 7-20

Horsepower Capacities of $\frac{3}{8}$"
   Diameter Braided Endless Round Belts 7-21

Horsepower Capacities of $\frac{9}{16}$"
   Diameter Wound Endless Round Belts 7-22

Method of Connection: Adjustable-Speed Drive 7-23

Method of Connection: Constant-Speed Drive 7-24

Maximum Horsepower: Two-Bearing Motors
   with Chain Drives 7-26

Horsepower Limits: Two-Bearing Motors with
   Belt Drive 7-26

Speed Limitations: Belt, Gear, and Chain Drives 7-27

CHAPTER 1
Formulas and Fundamentals

All electric motors operate by using electromagnetic induction, which is the interaction between conductors, currents, and magnetic fields. Any time an electrical current passes through a conductor (of which copper wires are the most common type), it causes a magnetic field to form around that conductor. Conversely, any time a magnetic field moves through a conductor, it induces (causes to flow) an electrical current in that conductor. Motors operate by making use of this, in combination with magnetic attraction and repulsion.

Electric motors turn electrical energy into mechanical movement.

When an electrical current flows through the motor's windings a strong magnetic field forms. This magnetic field attracts the rotor, and it moves it toward the magnetic field, causing the initial movement of the motor. This movement is continued by various means of rotating the magnetic field. The most common method of doing this is by using several different windings and sending current to them alternately, thus causing magnetic strength to be in one place one moment, and another place the next. The rotor will follow these fields, causing continuous motion.

While there are any number of variations to the above, all motors operate in this way.

INDUCTANCE

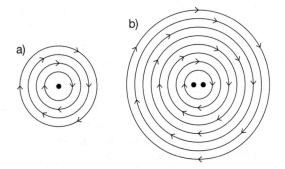

a)

b)

The magnetic field around one conductor (a)
and the combined magnetic fields of two
conductors (b). The conductors of both (a) and
(b) are carrying the same amount of amps.

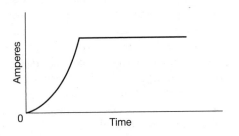

The delay in current flow because of
inductive reactance in a coil wire.

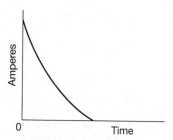

**Current flow maintained briefly because
of inductive reactance in a coil of wire
when the voltage is shut off.**

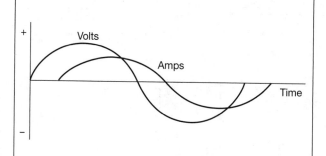

**Delay in current flow in an AC circuit by inductive
reactance in a coil of wire.**

CAPACITANCE

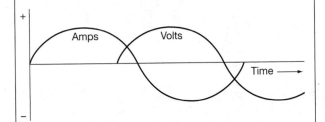

The leading current flow in an AC circuit caused by capacitive reactance.

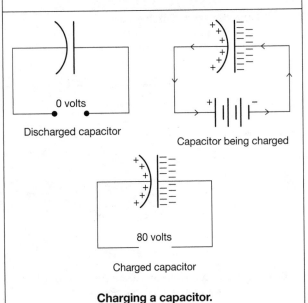

0 volts

Discharged capacitor

Capacitor being charged

80 volts

Charged capacitor

Charging a capacitor.

MECHANICAL AND ELECTRICAL DEGREES

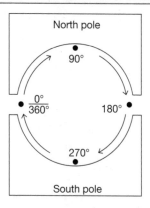

How electrical degrees compare with mechanical degrees in a two-pole motor.

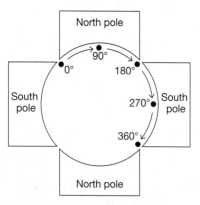

How 360 electrical degrees compare with 360 mechanical degrees in a four-pole motor.

POLES

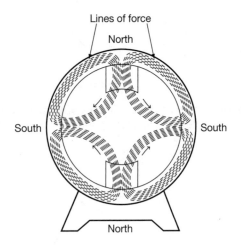

Lines of force

North

South

South

North

**If the two poles of a two-pole motor are
connected so that like polarity results,
two more poles will be formed by the lines
of force entering the frame.**

SYNCHRONOUS SPEED

Synchronous Speed (RPM) = $\dfrac{120\,f}{P}$

where f = frequency of supply

           P = the number of poles per phase for which the stator is wound

     RPM = revolutions per minute

      120 = constant

SLIP

The slip of an induction motor is the ratio of the difference between the rotating magnetic-field (synchronous) speed and the actual rotor speed.

Slip in RPM = synchronous speed − actual speed

Percent of RPM Slip =

$\dfrac{\text{synchronous speed} - \text{actual speed}}{\text{synchronous speed}} \times 100$

SPEED REGULATION

The speed regulation of a motor is the percentage drop in speed between no load and full load based on the full-load speed.

Percent Speed Regulation =

$\dfrac{\text{no-load speed} - \text{full-load speed}}{\text{full-load speed}} \times 100$

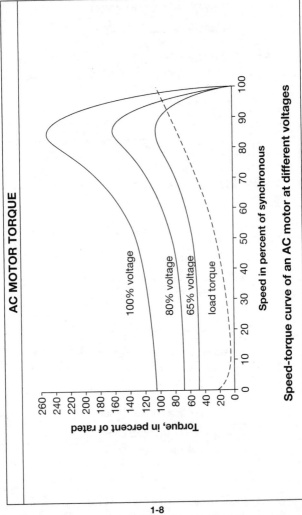

AC MOTOR TORQUE

Speed-torque curve of an AC motor at different voltages

Speed in percent of synchronous

Torque, in percent of rated

100% voltage

80% voltage

65% voltage

load torque

1-8

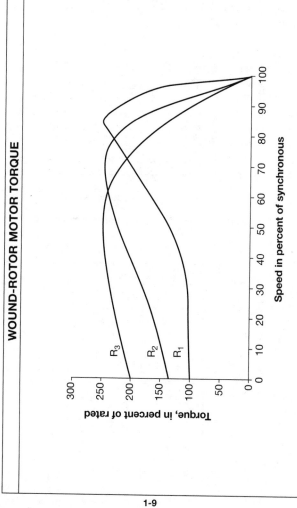

WOUND-ROTOR MOTOR TORQUE

R_3

R_2

R_1

Torque, in percent of rated

Speed in percent of synchronous

DC MOTOR TORQUE

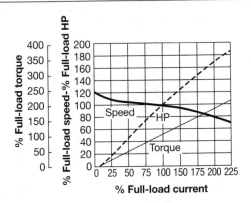

Typical characteristics for a compound motor

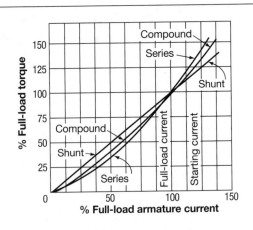

Torque characteristics for DC motors

SQUIRREL-CAGE MOTORS

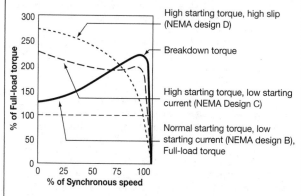

Squirrel-cage induction motors are classed by National Electrical Manufacturers Assn. (NEMA) according to locked-rotor torque, breakdown torque, slip, starting current, etc. Common types are Class B, C and D.

CLASS B: Most common type, has normal starting torque, low starting current. Locked-rotor torque (minimum torque at standstill and full voltage) is not less than 100% full-load for 2 and 4-pole motors, 200 HP and less; 40 to 75% for larger 2 pole motors; 50 to 125% for larger 4-pole motors.

CLASS C: Features high starting torque (locked-rotor over 200%), low starting current. Breakdown torque not less than 190% full-load torque. Slip at full load is between 1½ and 3%.

CLASS D: Have high slip, high starting torque, low starting current; are used on loads with high intermittent peaks. Driven machine usually has high-inertia flywheel. At no load motor has little slip; when peak load is applied, motor slip increases. Speed reduction lets driven machine absorb energy from flywheel rather than power line.

STARTING: Full-voltage, across-the-line starting is used where power supply permits and full-voltage torque and acceleration are not objectionable. Reduction in starting kVA cuts locked-rotor and accelerating torques.

WOUND-ROTOR MOTORS

The wound-rotor (slip-ring) induction motor's rotor winding connects through slip-rings to an external resistance that is cut in and out by a controller.

RESISTANCE vs TORQUE: Resistance of rotor winding affects torque developed at any speed. A high-resistance rotor gives high starting torque with low starting current. But low slip at full load, good efficiency and moderate rotor heating takes a low-resistance rotor. Left-hand curves show rotor-resistance effect on torque. With all resistance in, R1, full-load starting torque is developed at less than 150% full-load current. Successively shorting out steps, at standstill, develops about 225% full-load torque at R4. Cutting out more reduces standstill torque. Motor operates like a squirrel-cage motor when all resistance is shorted out.

SPEED CONTROL: Having resistance left in, decreases speed regulation. Righthand curves for a typical motor show that with only two steps shorted out, motor operates at 65% synchronous speed because motor torque equals load torque at that speed. But if load torque drops to 50%, motor shoots forward to about 65% synchronous speed.

However slip rings are normally shorted after motor comes up to speed. Or, for short-time peak loads, motor is operated with a step or two of resistance cut in. At light loads, motor runs near synchronous speed. When peak loads come on, speed drops; flywheel effect of motor and load cushions power supply from load peak.

OTHER FEATURES: In addition to high-starting-torque, low-starting-current applications, wound-rotor motors are used (1) for high-inertia loads where high slip losses that would have to be dissipated in the rotor of a squirrel-cage motor, in coming up to speed, can be given off as heat in wound-rotor's external resistance (2) where frequent starting, stopping and speed control are needed (3) for continuous operation at reduced speed. (Example: boiler draft fan – combustion control varies external resistance to adjust speed, damper regulates air flow between step speeds and below 50% speed.)

CONTROLS used are across-the-line starters with proper protection (fuses, breakers, etc.) air secondary control with 5 to 7 resistance steps.

SYNCHRONOUS MOTORS

Synchronous motors run at a fixed or synchronous speed determined by line frequency and number of poles in machine. (RPM=120 x frequency/number of poles.) Speed is kept constant by locking action of an externally excited DC field. Efficiency is 1 to 3% higher than that of same-size-and-speed induction or DC motors. Also synchronous motors can be operated at power factors from 1.0 down to 0.2 leading for plant power-factor correction. Standard ratings are 1.0 and 0.9 leading PF; machines rated down near 0.2 leading are called synchronous condensers.

STARTING: Pure synchronous motors are not self-starting; so in practice they are built with damper or amortisseur windings. With the field coil shorted through discharge resistor, damper winding acts like a squirrel-cage rotor to bring motor practically to synchronous speed; then field is applied and motor pulls into synchronism, providing motor has developed sufficient "pull-in" torque. Once in synchronism, motor keeps constant speed as long as load torque does not exceed maximum or "pull-out" torque; then machine drops out of synchronism. Driven machine is usually started without load. Low-speed motors may be direct connected.

FIELD AND PF: While motors are rated for specific power factors...at constant power, increasing DC field current causes power factor to lead, decreasing field tends to make PF lag. But either case increases copper losses.

TYPES: Polyphase synchronous motors in general use are: (1) high-speed motors, 500 RPM up, (a) general-purpose, 500 to 1800 RPM, 200 HP and below, (b) high-speed units over 200, HP including most 2-pole motors, (2) low-speed motors below 500 RPM, (3) special high-torque motors.

COSTS: Small high-speed synchronous motors cost more than comparable induction motors, but with low-speed and large high-speed motor, costs favor the synchronous motor. Cost of leading PF motors increases approximately inversely proportional to the decrease from unity power factor.

DC MOTORS

Chief reason for using DC motors, assuming normal power source is AC, lies in the wide and economical ranges possible of speed control and starting torques. But for constant-speed service, AC motors are generally preferred because they are more rugged and have lower first cost.

STARTING TORQUE: With a shunt motor, torque is proportional to armature current, because field flux remains practically constant for a given setting of the field rheostat. However, the flux of a series field is affected by the current through it. At light loads, flux varies directly with a current, so torque varies as the square of the current. The compound motor (usually cumulative) lies in between the shunt and series motors as to torque.

Upper limit of current input on starting is usually 1.5 to 2 times full-load current to avoid overheating the commutator, excessive fedder drops or peaking generator. Shunt-motor starting boxes usually allow 125% current at first notch. So motor can develop 125% starting torque. Series motors can develop higher starting torques at same current, since torque increases as current squared. Compound motors develop starting torques higher than shunt motors according to amount of compounding.

SPEED CONTROL: Shunt motor speeds drop only slightly (5% or less) from no load to full load. Decreasing field current raises speed; increasing field reduces speed. But speed is still practically constant for any one field setting. Speed can be controlled by resistance in the armature circuit but regulation is poor.

Series motor speeds decrease much more with increased load, and, conversely, begin to race at low loads, dangerously so if load is completely removed. Speed can be reduced by adding resistance into the armature circuit, increased by shunting the series filed with resistance or short-circuiting series turns.

Compound motors have less constant speed than shunt motors and can be controlled by shunt-field rheostat.

OHM'S LAW FOR ALTERNATING CURRENT

For the following Ohm's Law formulas for AC current, θ is the phase angle in degrees where current lags voltage (in inductive circuit) or by which current leads voltage (in a capacitive circuit). In a resonant circuit (such as 120VAC) the phase angle is 0° and Impedance = Resistance

$$\text{Current in amps} = \frac{\text{Voltage in volts}}{\text{Impedance in ohms}}$$

$$\text{Current in amps} = \sqrt{\frac{\text{Power in watts}}{\text{Impedance in ohms} \times \cos\theta}}$$

$$\text{Current in amps} = \frac{\text{Power in watts}}{\text{Voltage in volts} \times \cos\theta}$$

$$\text{Voltage in volts} = \text{Current in amps} \times \text{Impedance in ohms}$$

$$\text{Voltage in volts} = \frac{\text{Power in watts}}{\text{Current in amps} \times \cos\theta}$$

$$\text{Voltage in volts} = \sqrt{\frac{\text{Power in watts} \times \text{Impedance in ohms}}{\cos\theta}}$$

$$\text{Impedance in ohms} = \text{Voltage in volts} / \text{Current in amps}$$

$$\text{Impedance in ohms} = \text{Power in watts} / (\text{Current in amps}^2 \times \cos\theta)$$

$$\text{Impedance in ohms} = (\text{Voltage in volts}^2 \times \cos\theta) / \text{Power in watts}$$

$$\text{Power in watts} = \text{Current in amps}^2 \times \text{Impedance in ohms} \times \cos\theta$$

$$\text{Power in watts} = \text{Current in amps} \times \text{Voltage in volts} \times \cos\theta$$

$$\text{Power in watts} = \frac{(\text{Voltage in volts})^2 \times \cos\theta}{\text{Impedance in ohms}}$$

OHM'S LAW FOR DIRECT CURRENT

$$\text{Current in amps} = \frac{\text{Voltage in volts}}{\text{Resistance in ohms}} = \frac{\text{Power in watts}}{\text{Voltage in volts}}$$

$$\text{Current in amps} = \sqrt{\frac{\text{Power in watts}}{\text{Resistance in ohms}}}$$

Voltage in volts = Current in amps × Resistance in ohms

Voltage in volts = Power in watts / Current in amps

$$\text{Voltage in volts} = \sqrt{\text{Power in watts} \times \text{Resistance in ohms}}$$

Power in watts = (Current in amps)^2 × Resistance in ohms

Power in watts = Voltage in volts × Current in amps

Power in watts = (Voltage in volts)^2 / Resistance in ohms

Resistance in ohms = Voltage in volts / Current in amps

Resistance in ohms = Power in watts / (Current in amps)^2

POWER FACTOR

An AC electrical system carries two types of power: (1) true power, watts, that pulls the load (Note: Mechanical load reflects back into an AC system as resistance.) and (2) reactive power, vars, that generates magnetism within inductive equipment. The vector sum of these two will give actual volt-amperes flowing in the circuit (see diagram right). Power factor is the cosine of the angle between true power and volt-amperes.

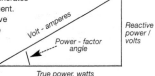

Volt - amperes

Reactive power / volts

Power - factor angle

True power, watts

SINGLE-PHASE POWER

Power of a single-phase AC circuit equals voltage times current times power factor:

$$P_{watts} = E_{volts} \times I_{amps} \times PF.$$

To figure reactive power, vars squared equals volt-amperes squared minus power squared, or

$$VARS = \sqrt{(VA)^2 - (P)^2}$$

THREE-PHASE AC CIRCUITS AND THE UTILIZATION OF POWER

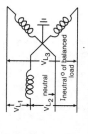

Sine waves are actually an oscillograph trace taken at any point in a three-phase system. (Each voltage or current wave actually comes from a separate wire but are shown for comparison on common base.) There are 120° between each voltage. At any instant the algebraic sum (measured up and down from centerline) of these three voltages is zero. When one voltage is zero, the other two are 86.6% maximum and have opposite signs.

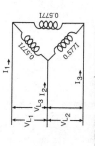

DELTA CONNECTION

Winding voltages equal line voltages, but currents split up so 0.577 I_{line} flows through windings.

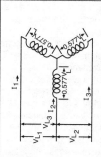

FOUR-WIRE SYSTEM

Most popular secondary distribution setup. V_1 is usually 208 V which feeds small power loads. Lighting loads at 120 V tap from any line to neutral.

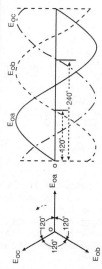

WYE CONNECTION

Consider the three windings as primary of transformer. Current in all windings equals line current, but volts across windings = 0.577 x line volts.

TYPES OF POWER

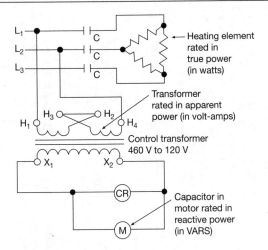

L₁

L₂

L₃

C

C

C

→ Heating element rated in true power (in watts)

Transformer rated in apparent power (in volt-amps)

H₃   H₂
H₁   H₄

Control transformer 460 V to 120 V

X₁   X₂

CR

M

Capacitor in motor rated in reactive power (in VARS)

Reactive power is supplied to a reactive load (capacitor/coil) and is measured in volt-amps reactive (VARS). The capacitor on a motor uses reactive power to keep the capacitor charged. The capacitor uses no true power because it performs no actual work such as producing heat or motion.

In an AC circuit containing only resistance, the power in the circuit is true power. However, almost all AC circuits include capacitive reactance (capacitors) and/or inductive reactance (coils). Inductive reactance is the most common, because all motors, transformers, solenoids and coils have inductive reactance.

Apparent power represents a load or circuit that includes both true power and reactive power and is expressed in volt-amps (VA), kilovolt amps (kVA) or megavolt amps (MVA) Apparent power is a measure of component or system capacity because apparent power considers circuit current regardless of how it is used. For this reason, transformers are sized in volt-amps rather than in watts.

TRUE POWER AND APPARENT POWER

True power is the actual power used in an electrical circuit and is expressed in watts (W). *Apparent power* is the product of voltage and current in a circuit calculated without considering the phase shift that may be present between total voltage and current in the circuit. Apparent power is measured in volt-amperes (VA). A phase shift exists in most AC circuits that contain devices causing capacitance or inductance.

True power equals apparent power in an electrical circuit containing only resistance. True power is less than apparent power in a circuit containing inductance or capacitance.

Capacitance is the property of a device that permits the storage of electrically separated charges when potential differences exist between the conductors. *Inductance* is the property of a circuit that causes it to oppose a change in current due to energy stored in a magnetic field; i.e: coils .

To calculate true power, apply the formula:

$$P_T = I^2 \times R$$

where

> P_T = true power (in watts)
>
> I = total circuit current (in amps)
>
> R = total resistive component of the circuit (in ohms)

To calculate apparent power, apply the formula:

$$P_A = E \times I$$

where

> P_A = apparent power (in volt-amps)
>
> E = measured voltage (in volts)
>
> I = measured current (in amps)

POWER FACTOR FORMULA

Power factor is the ratio of true power used in an AC circuit to apparent power delivered to the circuit.

$$PF = \frac{P_T}{P_A}$$

where

> PF = power factor (percentage)
>
> P_T = true power (in watts)
>
> P_A = apparent power (in volt-amps)

AC/DC POWER FORMULAS

To Find	For Direct Current	For Alternating Current		
		1φ, 115 or 120 V	1φ, 208, 230 or 240 V	3φ – ALL VOLTAGES
Amperes when Horsepower is known	$\dfrac{HP \times 746}{E \times E_{FF}}$	$\dfrac{HP \times 746}{E \times E_{FF} \times PF}$	$\dfrac{HP \times 746}{E \times E_{FF} \times PF}$	$\dfrac{HP \times 746}{1.73 \times E \times E_{FF} \times PF}$
Amperes when Kilowatts is known	$\dfrac{kW \times 1000}{E}$	$\dfrac{kW \times 1000}{E \times PF}$	$\dfrac{kW \times 1000}{E \times PF}$	$\dfrac{kW \times 1000}{1.73 \times E \times PF}$
Amperes when kVA is known		$\dfrac{kVA \times 1000}{E}$	$\dfrac{kVA \times 1000}{E}$	$\dfrac{kVA \times 1000}{1.73 \times E}$
Kilowatts	$\dfrac{I \times E}{1000}$	$\dfrac{I \times E \times PF}{1000}$	$\dfrac{I \times E \times PF}{1000}$	$\dfrac{I \times E \times 1.73 \times PF}{1000}$
Kilovolt-Amps		$\dfrac{I \times E}{1000}$	$\dfrac{I \times E}{1000}$	$\dfrac{I \times E \times 1.73}{1000}$
Horsepower (Output)	$\dfrac{I \times E \times E_{FF}}{746}$	$\dfrac{I \times E \times E_{FF} \times PF}{746}$	$\dfrac{I \times E \times E_{FF} \times PF}{746}$	$\dfrac{I \times E \times 1.73 \times E_{FF} \times PF}{746}$

HORSEPOWER FORMULAS

Current and Voltage Known	Speed and Torque Known

Current and Voltage Known

$$HP = \frac{E \times I \times E_{ff}}{746}$$

where
HP = horsepower
I = current (amps)
E = voltage (volts)
E_{ff} = efficiency
746 = constant

Speed and Torque Known

$$HP = \frac{rpm \times T}{5252}$$

where
HP = horsepower
rpm = revolutions per minute
T = torque (lb-ft)
5252 = constant

EFFICIENCY FORMULAS

Input and Output Power Known

$$E_{ff} = \frac{P_{out}}{P_{in}}$$

where
E_{ff} = efficiency (%)
P_{out} = output power (watts)
P_{in} = input power (watts)

Horsepower and Power Loss Known

$$E_{ff} = \frac{746 \times HP}{(746 \times HP) + W_l}$$

where
E_{ff} = efficiency (%)
746 = constant
HP = horsepower
W_l = watts lost

VOLTAGE UNBALANCE

$$V_u = \frac{V_d}{V_a} \times 100$$

where
V_u = voltage unbalance (%)
V_d = voltage deviation (volts)
V_a = voltage average (volts)
100 = constant

TEMPERATURE CONVERSIONS

Convert °C to °F

$$°F = (1.8 \times °C) + 32$$

Convert °F to °C

$$°C = \frac{(°F - 32)}{1.8}$$

VOLTAGE DROP FORMULAS

The NEC® recommends a maximum 3% voltage drop for either the branch circuit or the feeder.

Single-Phase:

$$VD = \frac{2 \times R \times I \times L}{CM}$$

Three-Phase:

$$VD = \frac{1.732 \times R \times I \times L}{CM}$$

VD = Volts (voltage drop of the circuit)

R = 12.9 Ohms/Copper or 21.2 Ohms/Aluminum (resistance constants for a conductor that is 1 circular mil in diameter and 1 foot long at an operating temperature of 75° C.)

I = Amps (load at 100 percent)

L = Feet (length of circuit from load to power supply)

CM = Circular-Mils (conductor wire size)

2 = Single-Phase Constant

1.732 = Three-Phase Constant

CONDUCTOR LENGTH/VOLTAGE DROP

Voltage drop can be reduced by limiting the length of the conductors.

Single-Phase:

$$L = \frac{CM \times VD}{2 \times R \times I}$$

Three-Phase:

$$L = \frac{CM \times VD}{1.732 \times R \times I}$$

CONDUCTOR SIZE/VOLTAGE DROP

Increase the size of the conductor to decrease the voltage drop of circuit (reduce its resistance).

Single-Phase:

$$CM = \frac{2 \times R \times I \times L}{VD}$$

Three-Phase:

$$CM = \frac{1.732 \times R \times I \times L}{VD}$$

FORMULAS FOR SINE WAVES

Frequency	Period	Peak-to-Peak Value
$f = \dfrac{1}{T}$ where f = frequency (in hertz) 1 = constant T = period (in seconds)	$T = \dfrac{1}{f}$ where f = frequency (in hertz) 1 = constant T = period (in seconds)	$V_{p\text{-}p} = 2 \times V_{max}$ where $V_{p\text{-}p}$ = peak-to-peak value 2 = constant V_{max} = peak value

Average Value	RMS Value
$V_{avg} = V_{max} \times .637$ where V_{avg} = average value (in volts) V_{max} = peak value (in volts) .637 = constant	$V_{rms} = V_{max} \times .707$ where V_{rms} = rms value (in volts) V_{max} = peak value (in volts) .707 = constant

CALCULATING ROOT-MEAN-SQUARE (RMS)

Effective (RMS) value = 0.707 x Peak value

Effective (RMS) value = 1.11 x Average value

Average value = 0.637 x Peak value

Average value = 0.9 x Effective (RMS) value

Peak value = 1.414 x Effective (RMS) value

Peak value = 1.57 x Average value

Peak-to-Peak value = 2 x Peak value

Peak-to-Peak value = 2.828 x Effective (RMS) value

SUMMARY OF SERIES, PARALLEL AND COMBINATION CIRCUITS

To Find	Series Circuits	Parallel Circuits	Series/Parallel
Resistance (R) Ohm Ω	$R_T = R_1 + R_2 + R_3$ Sum of individual resistances	$\frac{1}{R_T} = \frac{1}{R_1} + \frac{1}{R_2} + \frac{1}{R_3}$	Total resistance equals resistance of parallel portion and sum of series resistors
Current (I) Ampere A	$I_T = I_1 = I_2 = I_3$ The same throughout entire circuit	$I_T = I_1 + I_2 + I_3$ Sum of individual currents	Series rules apply to series portion of circuit Parallel rules apply to parallel part of circuit
Voltage (E) Volt V, E	$E_T = E_1 + E_2 + E_3$ Sum of individual voltages	$E_T = E_1 = E_2 = E_3$ Total voltage and branch voltage are the same	Total voltage is sum of voltage drops across each series resistor and each of the branches of parallel portion
Power (P) Watt W	$P_T = P_1 + P_2 + P_3$ Sum of individual wattages	$P_T = P_1 + P_2 + P_3$ Sum of individual wattages	$P_T = P_1 + P_2 + P_3$ Sum of individual wattages

1-24

CHAPTER 2
Motor Construction

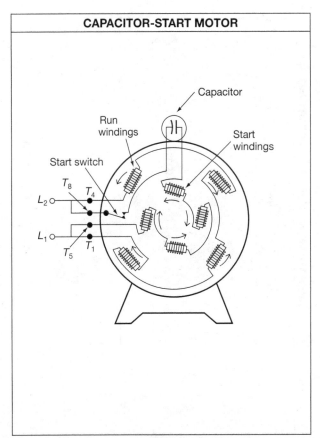

CAPACITOR-START MOTOR

Capacitor

Run windings

Start windings

Start switch

T_8
T_4
L_2
T_5
T_1
L_1

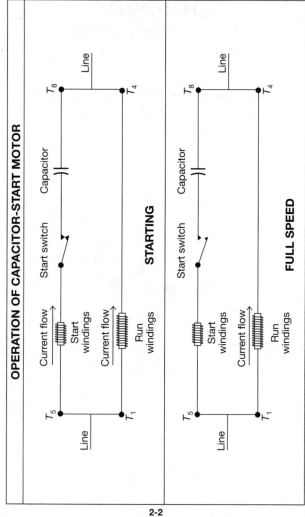

OPERATION OF CAPACITOR-START MOTOR

STARTING

Line — T_8

Capacitor

Start switch

Current flow → Start windings

Current flow → Run windings

T_5 — Line

T_1

T_4

FULL SPEED

Line — T_8

Capacitor

Start switch

Start windings

Current flow → Run windings

T_5 — Line

T_1

T_4

2-2

FOUR-POLE CAPACITOR-START MOTOR

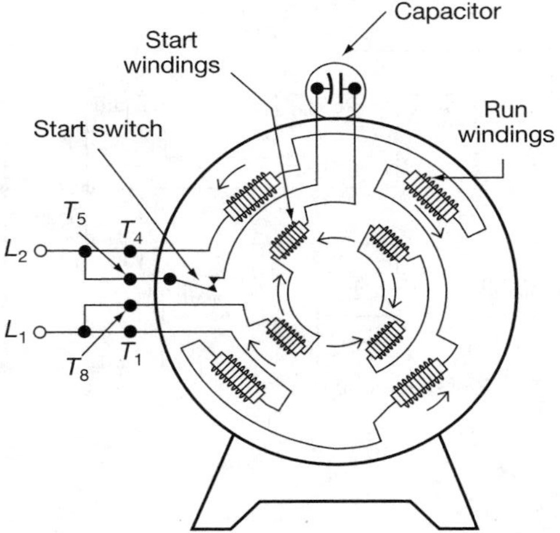

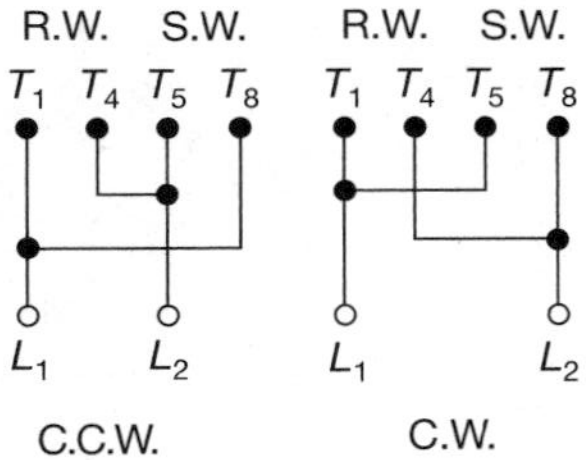

Capacitor

Run windings

Start windings

Start switch

C.C.W.

R.W.		S.W.	
T_1	T_4	T_5	T_8

L_1 L_2

C.W.

R.W.		S.W.	
T_1	T_4	T_5	T_8

L_1 L_2

Connect as above for desired rotation. To reverse – interchange T_5 and T_8.

2-3

FOUR-POLE, TWO-CIRCUIT CAPACITOR-START MOTOR

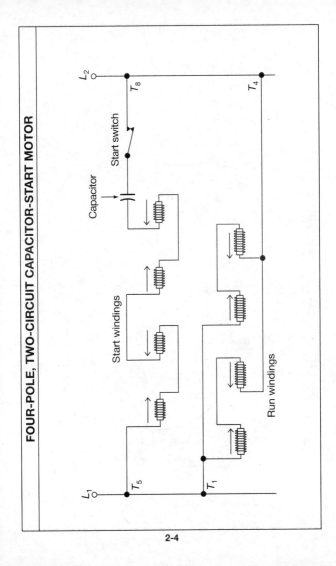

2-4

SIX-POLE CAPACITOR-START MOTOR

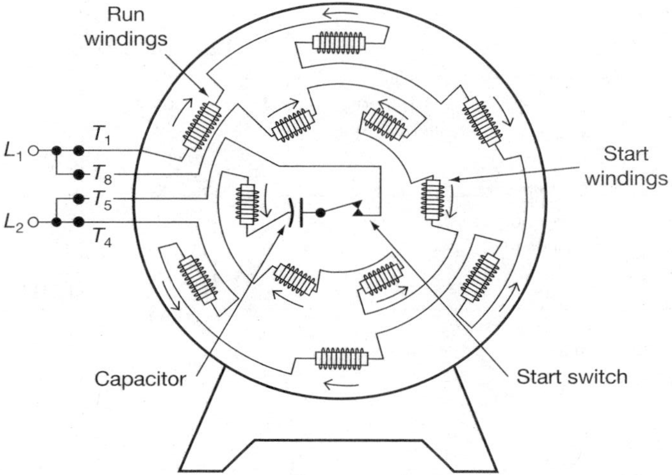

Start windings

Start switch

Run windings

Capacitor

T_1 T_8 T_5 T_4

L_1 L_2

2-5

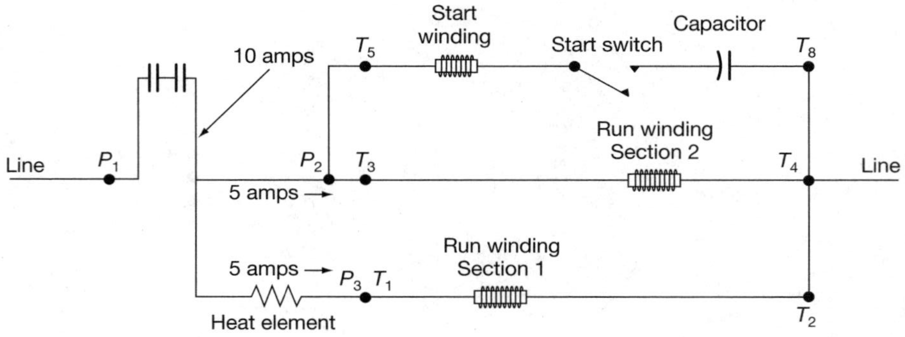

DUAL-VOLTAGE MOTORS

Line — T_8

Capacitor

Start switch

Start winding — T_5

Run winding Section 2

T_4 — Line

T_3

P_2 — 5 amps →

Run winding Section 1

T_2

P_3 T_1 — 5 amps →

Heat element

10 amps

P_1 — Line

Low Voltage

2-6

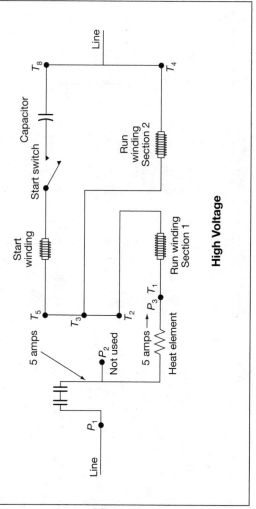

High Voltage

SINGLE-VOLTAGE CAPACITOR-START MOTOR

Current relay controls start winding.

2-8

DUAL-VOLTAGE CAPACITOR-START MOTOR

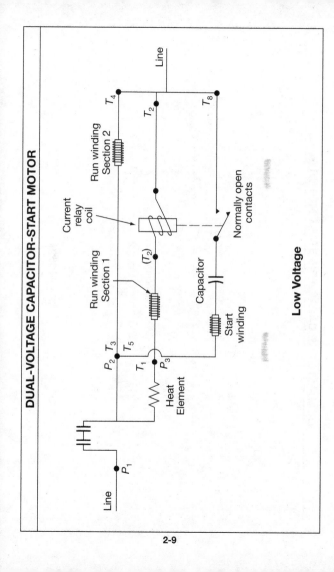

Low Voltage

2-9

DUAL-VOLTAGE CAPACITOR-START MOTOR (cont.)

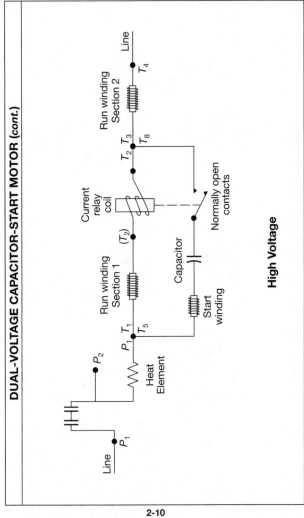

High Voltage

REVERSING SINGLE-PHASE MOTORS

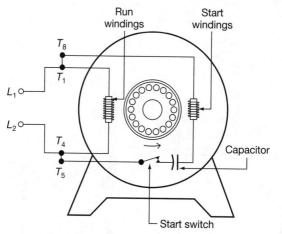

A capacitor-start motor with four leads.

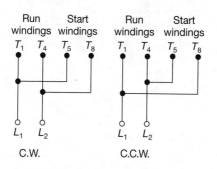

Connections for clockwise and
counterclockwise rotation.

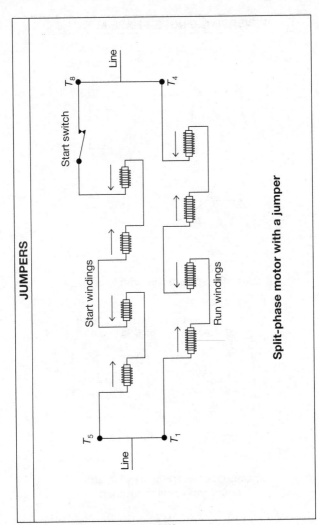

Split-phase motor with a jumper

JUMPERS (cont.)

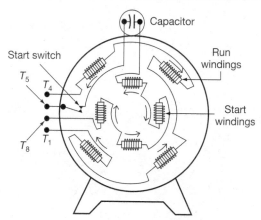

Capacitor

Start switch

Run windings

T_5 T_4

Start windings

T_8 T_1

One-circuit jumper connection

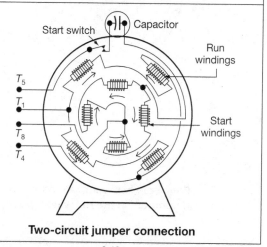

Start switch Capacitor

Run windings

T_5

T_1

Start windings

T_8

T_4

Two-circuit jumper connection

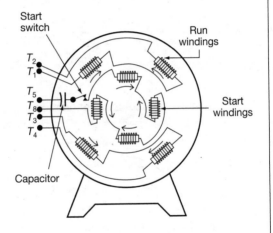

Start switch

Run windings

T_2
T_1

T_5
T_8
T_3
T_4

Start windings

Capacitor

**Four-pole two-voltage motor
with jumpers in the run windings.**

SIX-POLE MOTOR

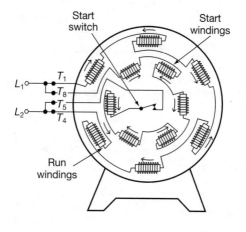

Start switch

Start windings

Run windings

L_1

T_1
T_8
T_5
T_4

L_2

2-15

REVERSING SPLIT-PHASE MOTORS

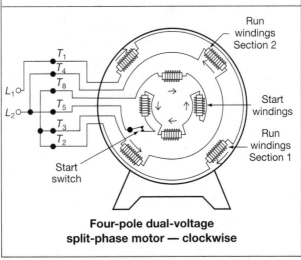

L_1 L_2
T_1 T_2 T_8 T_5 T_3 T_4

Run windings Section 1

Start windings

Start switch

Run windings Section 2

Four-pole dual-voltage split-phase motor — counterclockwise

L_1 L_2
T_1 T_4 T_8 T_5 T_3 T_2

Run windings Section 2

Start windings

Run windings Section 1

Start switch

Four-pole dual-voltage split-phase motor — clockwise

PERMANENT-SPLIT CAPACITOR MOTOR

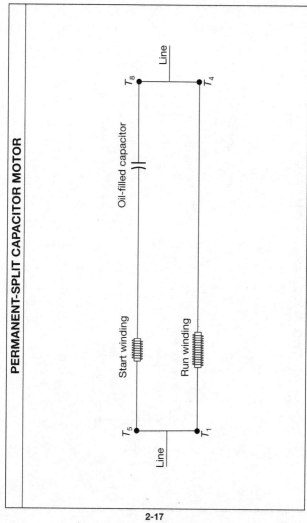

TWO-VALUE CAPACITOR MOTOR

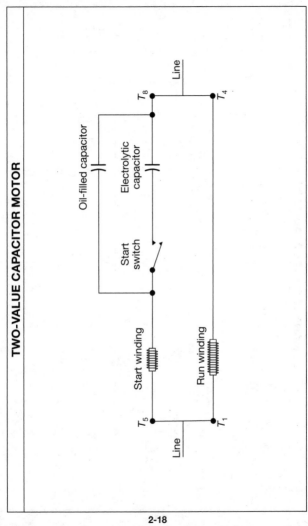

2-18

SYNCHRONOUS MOTORS

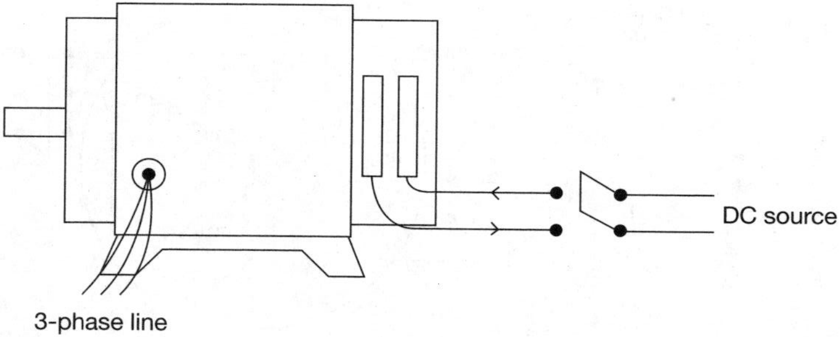

DC source
3-phase line

Basic synchronous motor connections

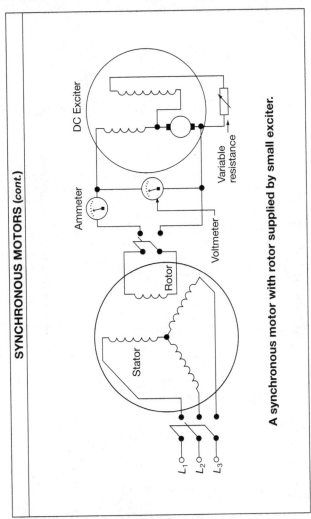

A synchronous motor with rotor supplied by small exciter.

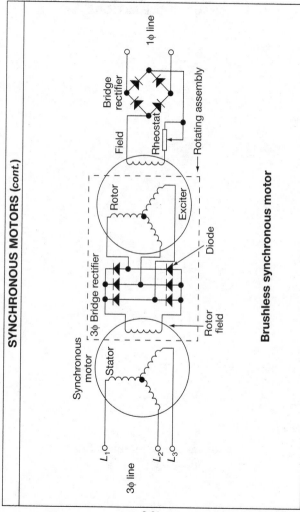

Brushless synchronous motor

SHADED-POLE MOTORS

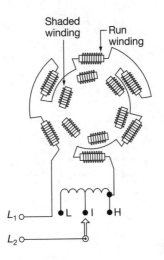

Shaded winding

Run winding

L_1

L_2

L I H

**Shaded-pole motor with
speed control by choke coil**

SHADED-POLE MOTORS (cont.)

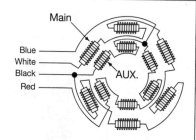

Three-speed shaded-pole motor

	L_1	L_2	Open
High speed	White	Black	Red, Blue
Int. speed	White	Blue	Red, Black
Low speed	White	Red	Blue, Black

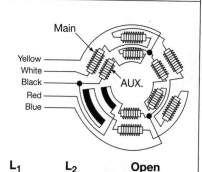

Four-speed shaded-pole motor

	L_1	L_2	Open
High speed	White	Black	Red, Blue, Yellow
Int. 2 speed	White	Yellow	Red, Blue, Black
Int. 3 speed	White	Blue	Red, Yellow, Black
Low speed	White	Red	Blue, Yellow, Black

SHADED-POLE MOTORS (cont.)

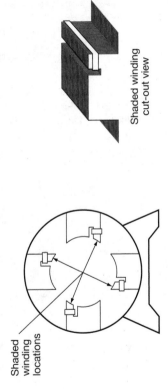

Shaded winding cut-out view

Shaded winding locations

A four-pole, shaded-pole motor showing field poles and shading windings.

SHADED-POLE MOTORS (cont.)

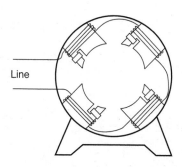

Line

A four-pole, shaded-pole motor with field poles connected in series for alternate polarity.

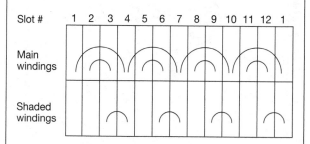

Slot #   1   2   3   4   5   6   7   8   9   10 11 12   1

Main windings

Shaded windings

The windings of a four-pole, 12-slot, distributed shaded-pole motor.

SHADED-POLE MOTORS (cont.)

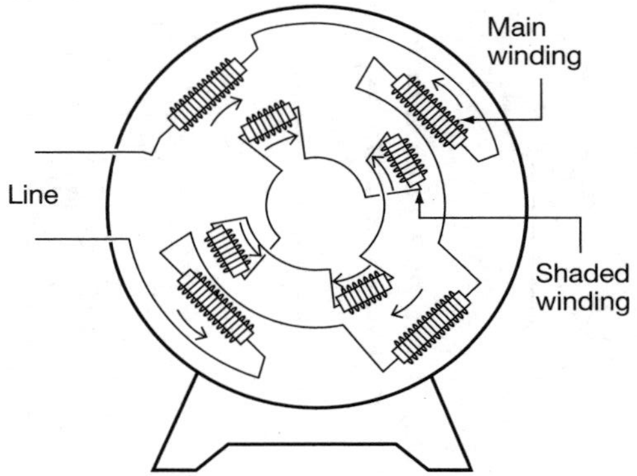

Main winding

Shaded winding

Line

A diagram of four-pole, distributed shaded-pole windings

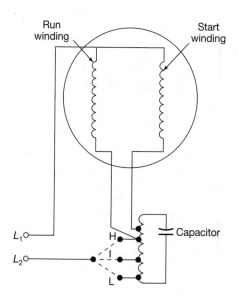

Basic capacitor motor

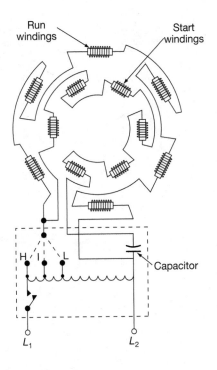

Three-speed motor

MULTI-SPEED SINGLE-PHASE MOTORS

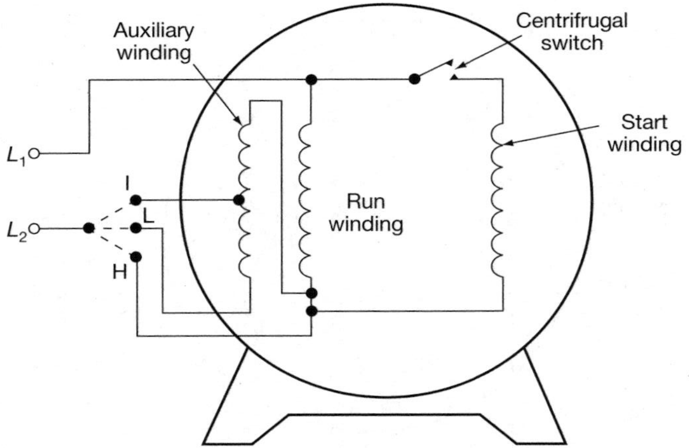

Centrifrugal switch

Start winding

Run winding

Auxiliary winding

L_1

L_2

I

L

H

Three-speed split-phase motor

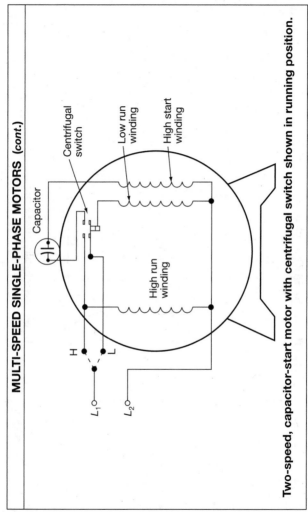

MULTI-SPEED SINGLE-PHASE MOTORS (cont.)

Capacitor

Centrifugal switch

Low run winding

High start winding

High run winding

H L

L_1

L_2

Two-speed, capacitor-start motor with centrifugal switch shown in running position.

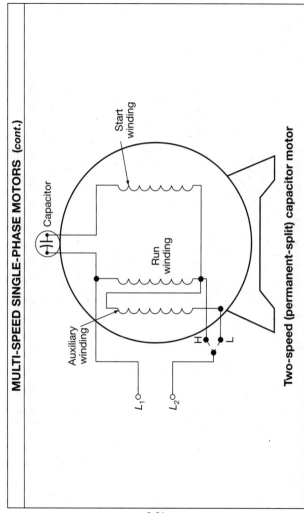

Two-speed (permanent-split) capacitor motor

Centrifugal switch

Low run winding

High start winding

High run winding

H   L

L_1

L_2

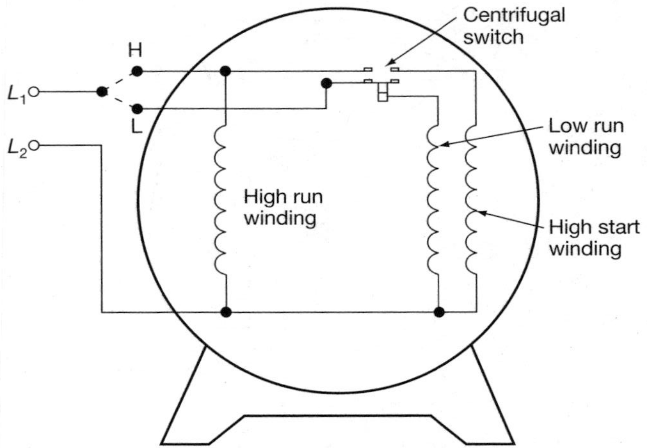

A two-speed, split-phase motor with two run windings and one start winding. Centrifugal switch shown in running position.

MULTI-SPEED SINGLE-PHASE MOTORS (cont.)

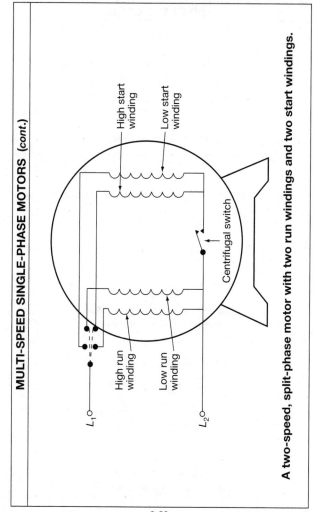

High start winding

Low start winding

Centrifugal switch

High run winding

Low run winding

L_1

L_2

A two-speed, split-phase motor with two run windings and two start windings.

UNIVERSAL MOTORS

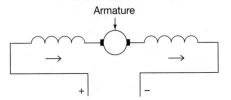

Basic universal motor

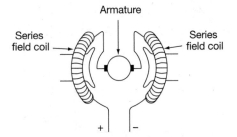

A universal motor connected for clockwise rotation.

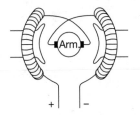

A universal motor connected for counterclockwise rotation.

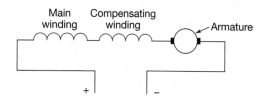

Compensated universal motor.

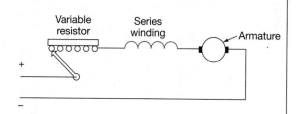

A universal motor speed controlled with a variable resistor in series with the motor.

UNIVERSAL MOTORS (cont.)

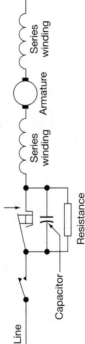

Pressure separates the carbons and decreases the current through the motor to vary speed

Speed control of a universal motor by a variation in contact resistance between carbon blocks.

UNIVERSAL MOTORS (cont.)

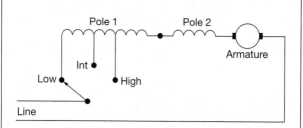

**Three speeds obtained in a universal
motor by tapping one field pole.**

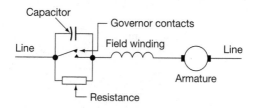

**Speed control of a universal
motor utilizing a centrifugal governor.**

UNIVERSAL MOTORS (cont.)

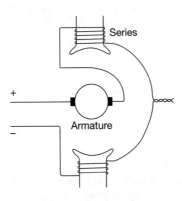

**A series connection of
a universal motor.**

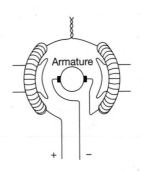

**A series connection
showing tapped field coils.**

CONNECTIONS FOR MULTI-SPEED SQUIRREL-CAGE MOTORS

Two Speeds — One Winding
Constant Horsepower

Speed	L_1	L_2	L_3	Open	Together
1 Low	T_1	T_2	T_3	—	T_4 T_5 T_6
2 High	T_6	T_4	T_5	T_1 T_2 T_3	—

Two Speeds — One Winding (cont.)
Constant Torque

Speed	L_1	L_2	L_3	Open	Together
1 Low	T_1	T_2	T_3	All others	—
2 High	T_6	T_4	T_5	—	T_1 T_2 T_3

CONNECTIONS FOR MULTI-SPEED SQUIRREL-CAGE MOTORS (cont.)

Two Speeds — One Winding (cont.)
Variable Torque

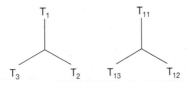

Speed	L_1	L_2	L_3	Open	Together
1 Low	T_1	T_2	T_3	All others	—
2 High	T_6	T_4	T_5	—	T_1 T_2 T_3

Two Speeds — Two Windings
Constant Torque, Variable Torque or Constant Horsepower

Speed	L_1	L_2	L_3	Open
1 Low	T_1	T_2	T_3	T_{11} T_{12} T_{13}
2 High	T_{11}	T_{12}	T_{13}	T_1 T_2 T_3

CONNECTIONS FOR MULTI-SPEED SQUIRREL-CAGE MOTORS (*cont.*)

Two Speeds — Two Windings (*cont.*)
Constant Torque, Variable Torque or Constant Horsepower

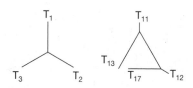

Speed	L₁	L₂	L₃	Open
1 Low	T_1	T_2	T_3	T_{11} T_{12} T_{13} T_{17}
2 High	T_{11}	T_{12}	T_{13} T_{17}	T_1 T_2 T_3

Two Speeds — Two Windings (*cont.*)
Constant Torque, Variable Torque or Constant Horsepower

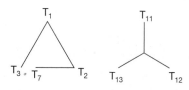

Speed	L₁	L₂	L₃	Open
1 Low	T_1	T_2	T_3 T_7	T_{11} T_{12} T_{13}
2 High	T_{11}	T_{12}	T_{13}	T_1 T_2 T_3 T_7

CONNECTIONS FOR MULTI-SPEED SQUIRREL-CAGE MOTORS (cont.)

Two Speeds — Two Windings (cont.)
Constant Torque

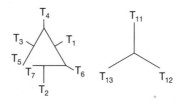

Speed	L_1	L_2	L_3	Open	Together
1 Low	T_1	T_2	T_3 T_7	All others	—
2 Int	T_6	T_4	T_5	All others	T_1 T_2 T_3 T_7
3 High	T_{11}	T_{12}	T_{13}	All others	—

Three Speeds — Two Windings
Constant Horsepower

Speed	L_1	L_2	L_3	Open	Together
1 Low	T_1	T_2	T_3	All others	T_4 T_5 T_6 T_7
2 Int	T_6	T_4	T_5 T_7	All others	—
3 High	T_{11}	T_{12}	T_{13}	All others	—

CONNECTIONS FOR MULTI-SPEED SQUIRREL-CAGE MOTORS (*cont.*)

Three Speeds — Two Windings (*cont.*)
Constant Horsepower

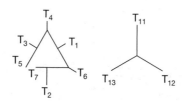

Speed	L_1	L_2	L_3	Open	Together
1 Low	T_1	T_2	T_3	All others	T_4 T_5 T_6 T_7
2 Int	T_{11}	T_{12}	T_{13}	All others	—
3 High	T_6	T_4	T_5 T_7	All others	—

Three Speeds — Two Windings (*cont.*)
Constant Horsepower

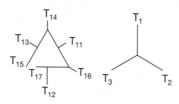

Speed	L_1	L_2	L_3	Open	Together
1 Low	T_1	T_2	T_3	All others	—
2 Int	T_{11}	T_{12}	T_{13}	All others	T_{14} T_{15} T_{16} T_{17}
3 High	T_{16}	T_{14}	T_{15} T_{17}	All others	—

WYE THREE-PHASE MOTORS

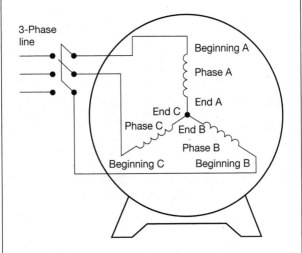

A star connection, also called Y connection.

WYE THREE-PHASE MOTORS (*cont.*)

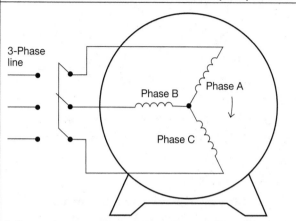

A three-phase motor connected to a three-phase line.

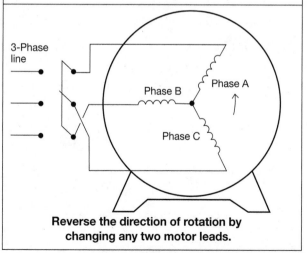

**Reverse the direction of rotation by
changing any two motor leads.**

WYE THREE-PHASE MOTORS (*cont.*)

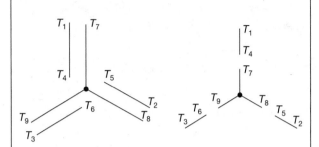

Standard markings for Y connected, dual-voltage motors

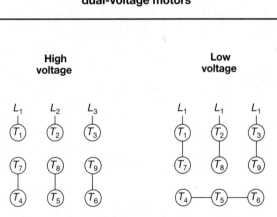

Y connections for dual-voltage

WYE THREE-PHASE MOTORS (cont.)

Voltage	L_1	L_2	L_3	Splice		
Low	T_1 T_7	T_2 T_8	T_3 T_9		T_4 T_5 T_6	
High	T_1	T_2	T_3	T_4 T_7	T_5 T_8	T_6 T_9

Markings and connections for a Y connected dual-voltage motor.

WYE THREE-PHASE MOTORS (*cont.*)

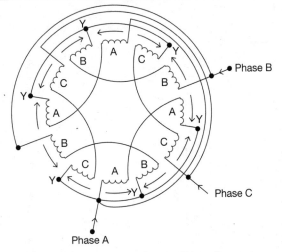

Four-pole, two-wye, three-phase diagram

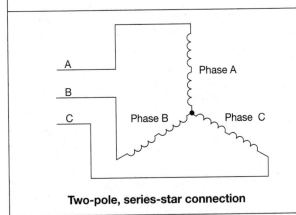

Two-pole, series-star connection

Phase A

Phase C

Phase B

A → B → C → A → B → C →

Y Y Y

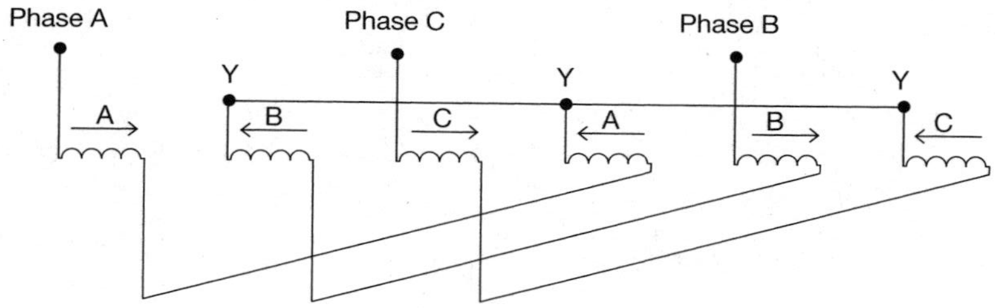

A diagram of a series-star connection

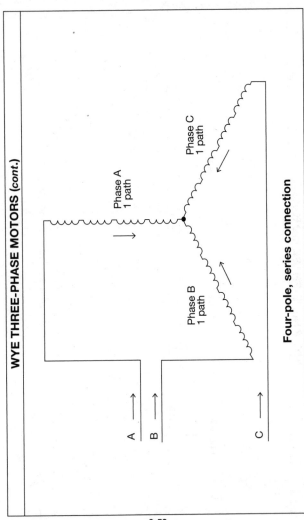

WYE THREE-PHASE MOTORS *(cont.)*

Phase A
1 path

Phase B
1 path

Phase C
1 path

A

B

C

Four-pole, series connection

2-50

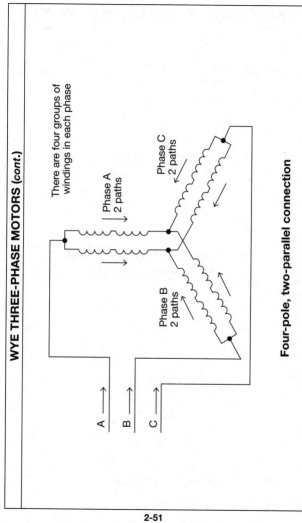

WYE THREE-PHASE MOTORS *(cont.)*

There are four groups of windings in each phase

Phase A
2 paths

Phase C
2 paths

Phase B
2 paths

A →

B →

C →

Four-pole, two-parallel connection

WYE-WOUND MOTOR FOR USE ON 240/480 VOLTS

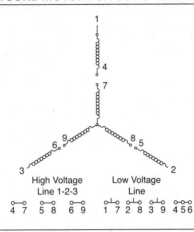

High Voltage
Line 1-2-3
4 7   5 8   6 9

Low Voltage
Line
1 7   2 8   3 9   4 5 6

NUMBERING OF DUAL-VOLTAGE WYE-WOUND MOTOR

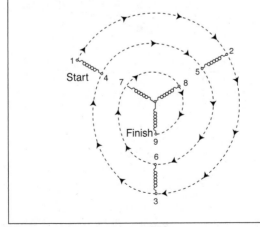

DELTA THREE-PHASE MOTORS

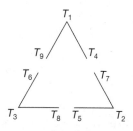

**Standard markings for
delta-connected dual-voltage motors**

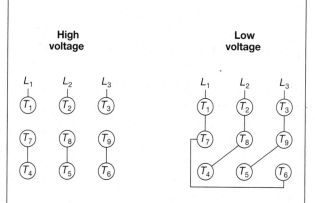

Delta connections for dual-voltage

2-53

DELTA THREE-PHASE MOTORS (cont.)

Voltage	L_1	L_2	L_3	Splice		
Low	$T_1 T_6 T_7$	$T_2 T_4 T_8$	$T_3 T_5 T_9$			
High	T_1	T_2	T_3	$T_4 T_7$	$T_5 T_8$	$T_6 T_9$

Markings and connections for delta-connected dual-voltage motors.

DELTA THREE-PHASE MOTORS (cont.)

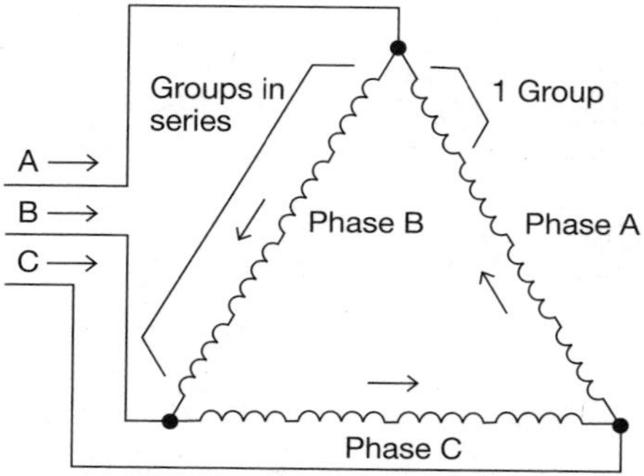

1 Group

Phase A

Phase B

Phase C

Groups in series

A →
B →
C →

Three-phase, four-pole, series delta motor

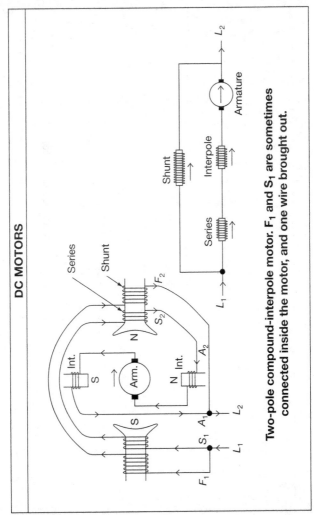

Two-pole compound-interpole motor. F₁ and S₁ are sometimes connected inside the motor, and one wire brought out.

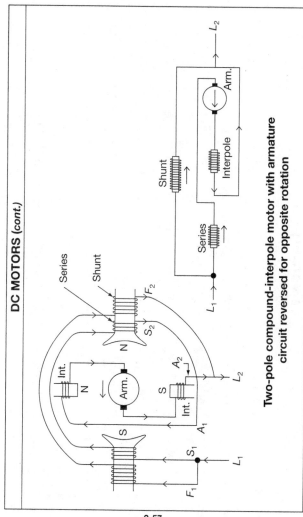

Two-pole compound-interpole motor with armature circuit reversed for opposite rotation

2-57

DC MOTORS (cont.)

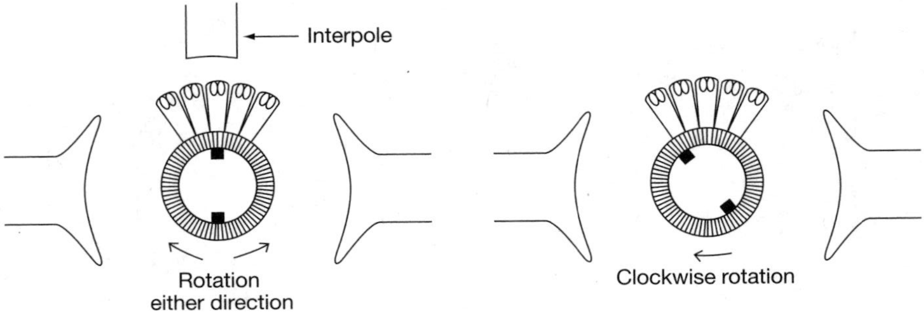

Interpole

Clockwise rotation

Rotation either direction

Brush positions for interpole and non-interpole motors.

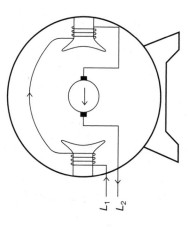

An equal amount of current flows through all elements of a series motor.

L_1
L_2

DC MOTORS (cont.)

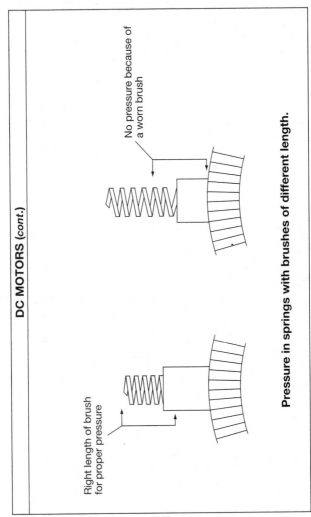

Right length of brush for proper pressure

No pressure because of a worn brush

Pressure in springs with brushes of different length.

DC MOTORS (cont.)

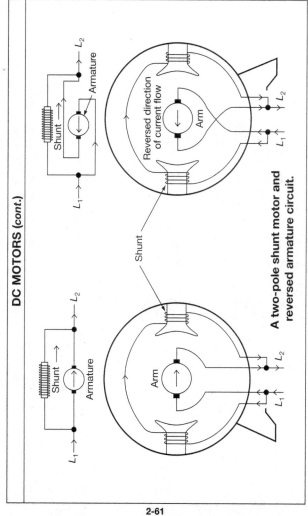

A two-pole shunt motor and reversed armature circuit.

DC MOTORS (cont.)

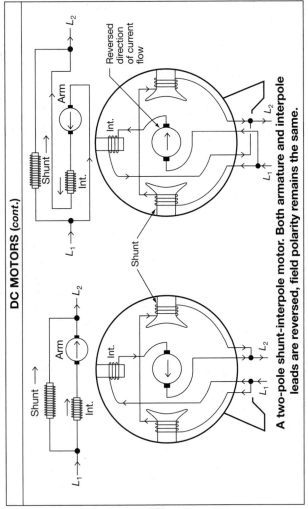

A two-pole shunt-interpole motor. Both armature and interpole leads are reversed, field polarity remains the same.

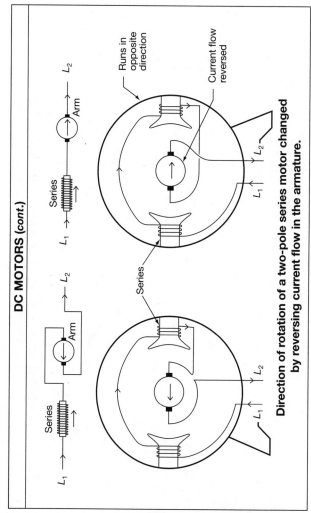

Runs in opposite direction

Current flow reversed

Series

Series

Arm

Series

Arm

Series

L_1

L_2

L_2

L_1

L_2

L_1

L_2

Direction of rotation of a two-pole series motor changed by reversing current flow in the armature.

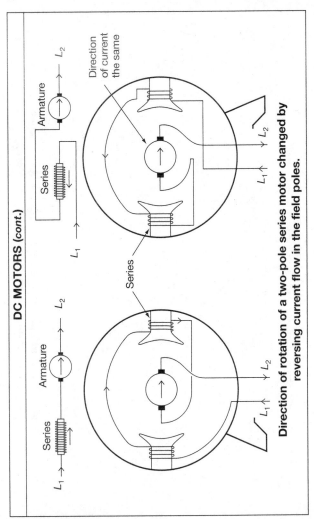

Direction of current the same

Armature

Series

L_2

L_1

Series

Armature

Series

L_2

L_1

L_1

L_2

Series

Direction of rotation of a two-pole series motor changed by reversing current flow in the field poles.

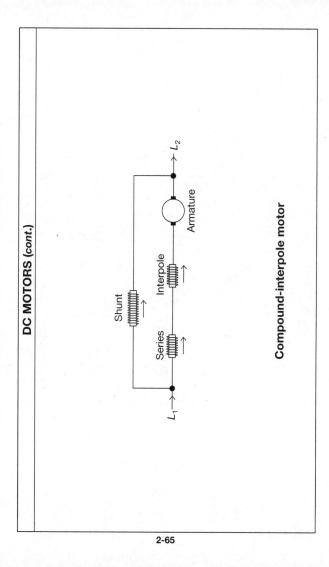

Compound-interpole motor

A two-pole compound-interpole motor with counterclockwise rotation.

Labels: Interpole polarity same as main pole; Series; Shunt; Interpole connected in series with armature; Int.; N; Arm; S; Int.; S; L_1; L_2

A two-pole compound-interpole motor with
one interpole connected in series with the armature.

DC MOTORS (cont.)

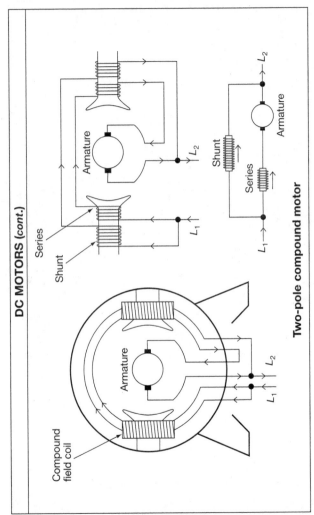

Two-pole compound motor

DC MOTORS (cont.)

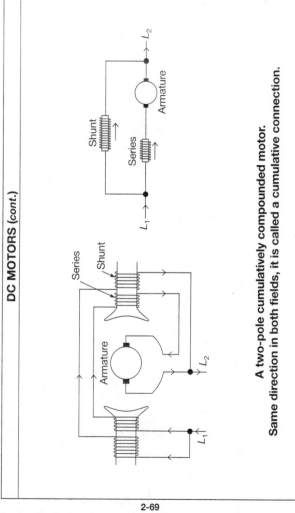

A two-pole cumulatively compounded motor.
Same direction in both fields, it is called a cumulative connection.

DC MOTORS (*cont.*)

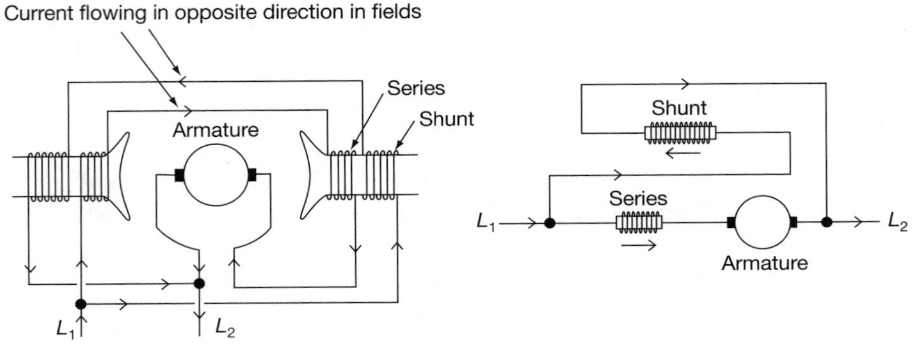
Current flowing in opposite direction in fields
Series
Shunt
Armature
L_1
L_2
Shunt
Series
Armature
L_1
L_2

Differentially connected compound motor

CHAPTER 3
Connections and Diagrams

DELTA-WOUND MOTOR FOR USE ON 240/480 VOLTS

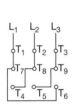

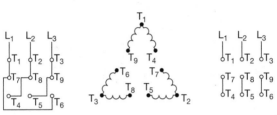

Low Voltage

High Voltage

WYE-WOUND MOTOR FOR USE ON 240/480 VOLTS

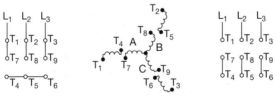

Low Voltage

High Voltage

SPLIT-PHASE MOTORS

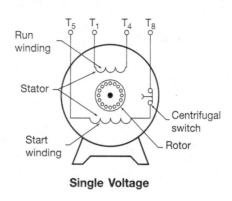

T_5 T_1 T_4 T_8

Run winding

Stator

Start winding

Centrifugal switch

Rotor

Single Voltage

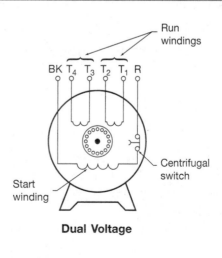

Run windings

BK T_4 T_3 T_2 T_1 R

Start winding

Centrifugal switch

Dual Voltage

SPLIT-PHASE MOTOR ROTATION

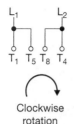
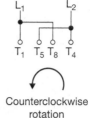

Clockwise rotation

Counterclockwise rotation

Single Voltage

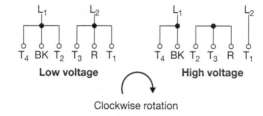

Clockwise rotation

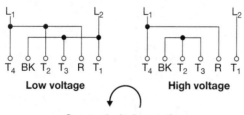

Counterclockwise rotation

Dual Voltage

SINGLE-VOLTAGE, 3φ, WYE-CONNECTED MOTOR

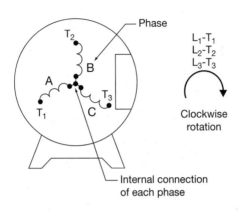

Phase

L_1-T_1
L_2-T_2
L_3-T_3

Clockwise rotation

Internal connection of each phase

SINGLE-VOLTAGE, 3φ, DELTA-CONNECTED MOTOR

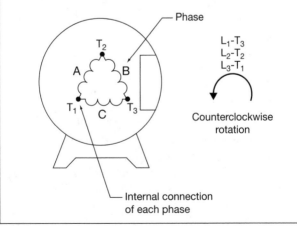

Phase

L_1-T_3
L_2-T_2
L_3-T_1

Counterclockwise rotation

Internal connection of each phase

3-4

DUAL-VOLTAGE, 3φ, WYE-CONNECTED MOTORS

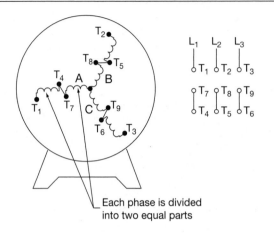

Each phase is divided into two equal parts

High Voltage (series)

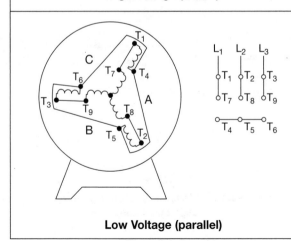

Low Voltage (parallel)

3-5

TWO-PHASE, AC MOTORS

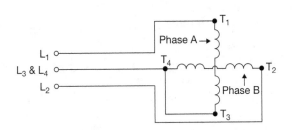

3-Wire

To reverse direction, interchange L_1 and L_2.

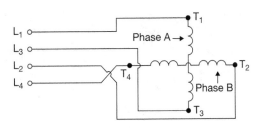

4-Wire

To reverse direction, interchange the
leads in one phase.

CONNECTIONS FOR A TWO-SPEED, CONSTANT HORSEPOWER, ONE WINDING MOTOR

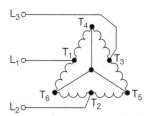

Low Speed

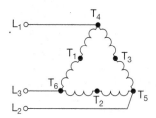

High Speed

Note: Torque decreases in the same ratio as speed increases maintaining constant horsepower.

CONNECTIONS FOR A TWO-SPEED, CONSTANT TORQUE, ONE WINDING MOTOR

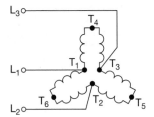

Low Speed

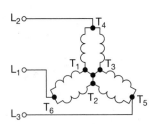

High Speed

Note: In a constant torque motor, the horse-power changes proportionally to the speed.

CONNECTIONS FOR A TWO-SPEED, VARIABLE TORQUE, ONE WINDING MOTOR

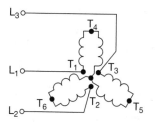

Low Speed

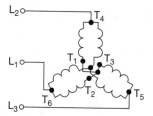

High Speed

Note: In a variable torque motor, the torque and the horsepower vary inversely with the speed.

CAPACITOR-START-CAPACITOR-RUN MOTOR

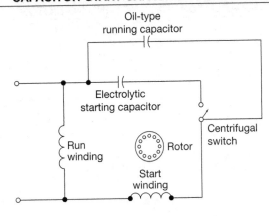

REVERSING A CAPACITOR-START MOTOR

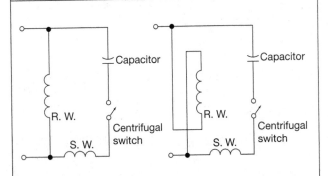

The reversal of a capacitor-start motor is accomplished by reversing the leads of the running winding.

CAPACITOR-START MOTOR VOLTAGE CONNECTIONS

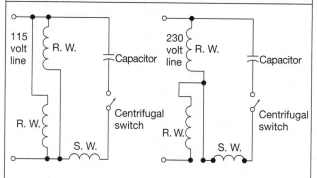

Capacitor-start motor, first connected across 115 volts and then across 230 volts.

TWO-SPEED CAPACITOR MOTOR

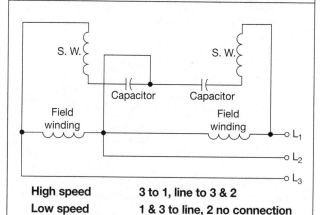

High speed   3 to 1, line to 3 & 2

Low speed   1 & 3 to line, 2 no connection

WOUND-ROTOR MOTOR SCHEMATIC

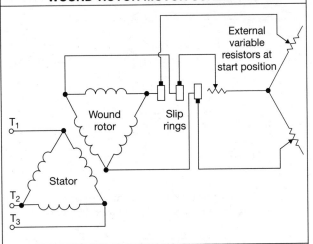

REVERSING SPLIT-PHASE MOTORS

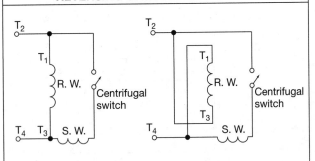

The reversal of a split-phase motor is accomplished by reversing the running winding leads.

TYPICAL MOTOR STARTER DIAGRAM

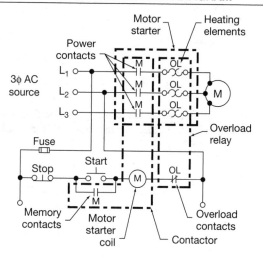

STEP-DOWN TRANSFORMER MOTOR CONTROL

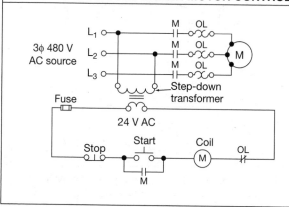

3-13

MOTOR CONTROL CIRCUITS

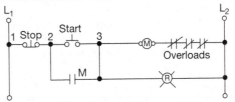

Magnetic starter with one stop-start station and a pilot lamp which burns to indicate that the motor is running.

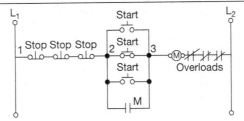

Magnetic starter with three stop-start stations.

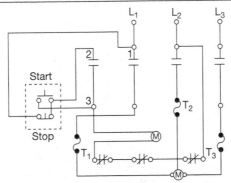

Diagrammatic representation of a magnetic three-phase starter with one start-stop station.

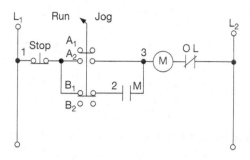

Jogging using a selector push button.

For "**Run**," the selector switch engages only A.
For "**Jog**," the selector switch engages both A and B.

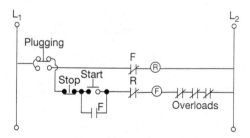

Magnetic starter with a plugging switch.

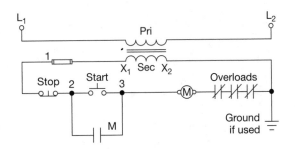

Low-voltage motor control using a transformer.

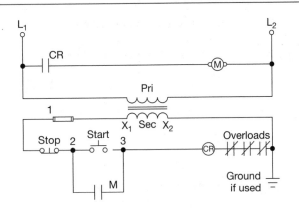

**Low-voltage motor control
using a transformer and control relay.**

MOTOR CONTROL CIRCUITS (cont.)

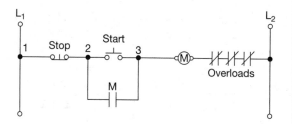

**Magnetic three-phase starter
with one start-stop station.**

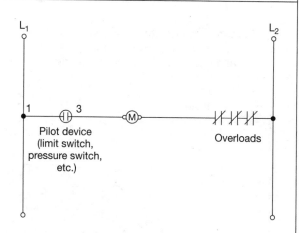

Maintained-contact control.

MOTOR CONTROL CIRCUITS *(cont.)*

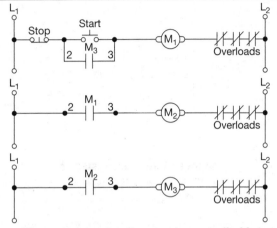

Three motors simultaneously controlled by
one stop-start station; if one overload device trips,
all three motors will stop running.

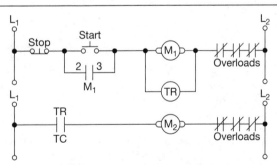

Two magnetic starters controlled by one
start-stop station; a starting-time delay
device between the two motors.

MOTOR CONTROL CIRCUITS *(cont.)*

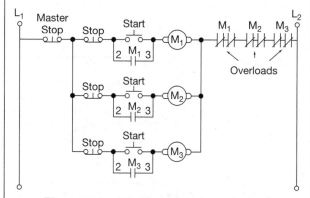

**Three separately started motors stopped
by one master stop station or stopped if
one overload device is tripped.**

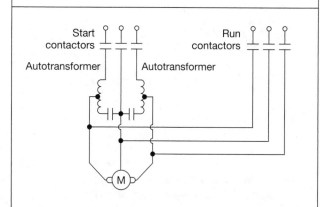

Starting compensator, with start and run contacts.

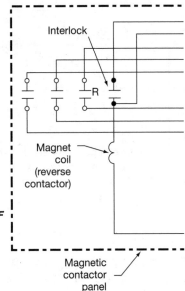

Interlock

R

Magnet
coil
(reverse
contactor)

Note:
Contactors R and F
are mechanically
interlocked

Magnetic
contactor
panel

**An across-the-line reversing type
starter controlling a three-phase motor.**

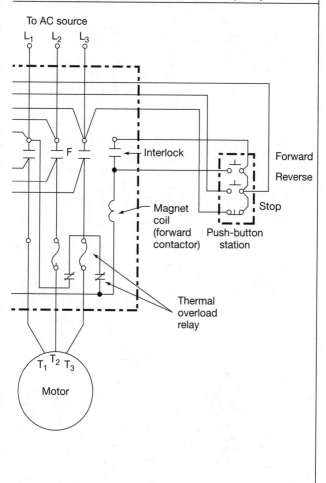

To AC source

L_1 L_2 L_3

F

Interlock

Forward

Reverse

Stop

Magnet coil (forward contactor)

Push-button station

Thermal overload relay

T_1 T_2 T_3

Motor

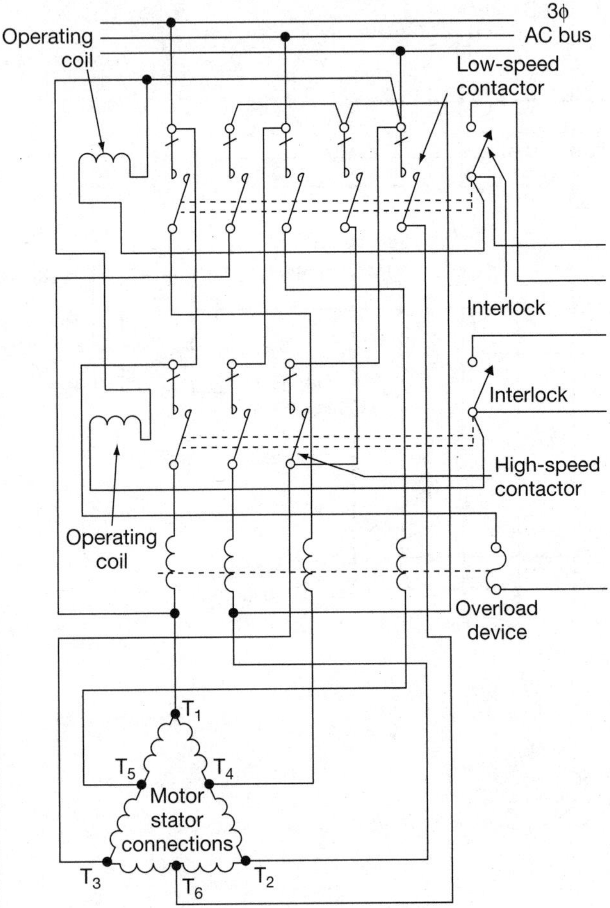

Operating coil

3φ AC bus

Low-speed contactor

Interlock

Interlock

High-speed contactor

Operating coil

Overload device

T_1

T_5 T_4

Motor stator connections

T_3 T_6 T_2

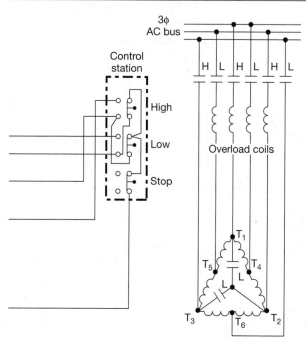

A two-speed AC motor with push-button control.

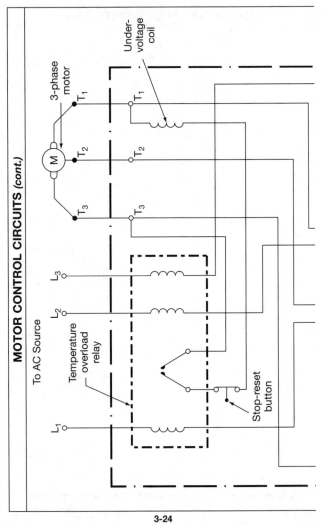

MOTOR CONTROL CIRCUITS (cont.)

To AC Source

3-phase motor

Under-voltage coil

Temperature overload relay

Stop-reset button

3-24

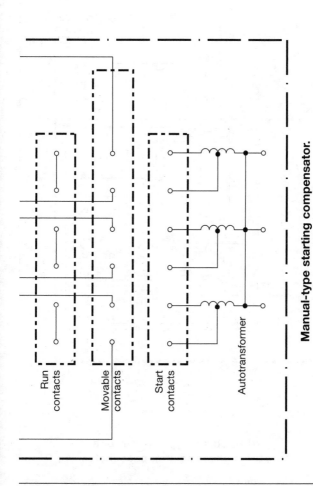

Manual-type starting compensator.

Run contacts

Movable contacts

Start contacts

Autotransformer

3-25

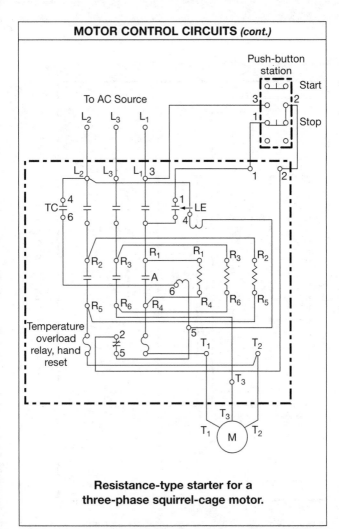

MOTOR CONTROL CIRCUITS (cont.)

Resistance-type starter for a
three-phase squirrel-cage motor.

3-26

MOTOR CONTROL CIRCUITS *(cont.)*

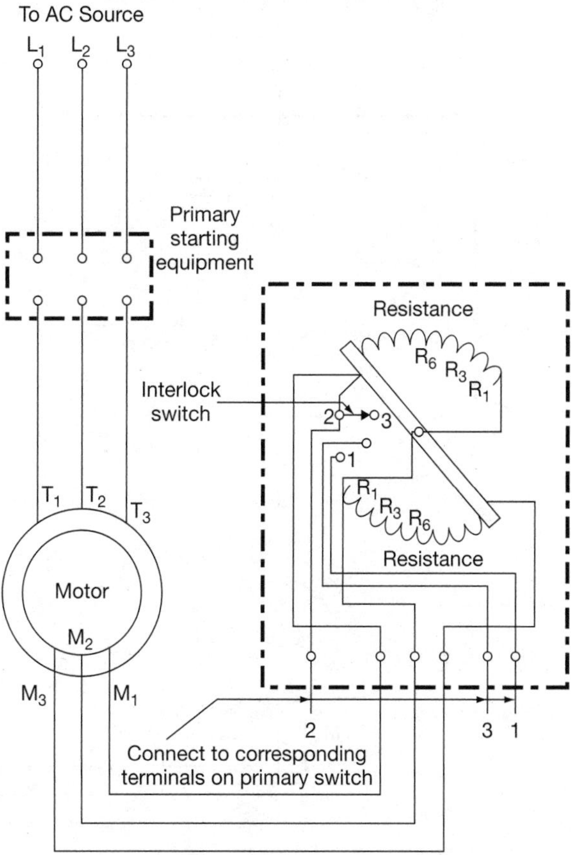

To AC Source
L1
L2
L3
Primary
starting
equipment
Resistance
R6 R3 R1
Interlock
switch
2
3
1
R1
R3
R6
Resistance
T1
T2
T3
Motor
M2
M3
M1
2
3 1
Connect to corresponding
terminals on primary switch

Diagram of a faceplate starter for a wound-rotor induction motor.

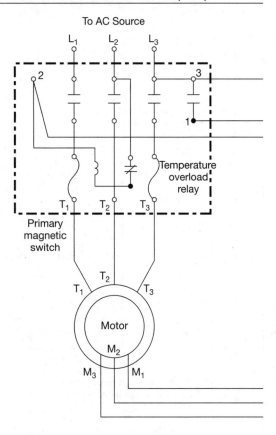

To AC Source

Secondary speed-regulating rheostat and a

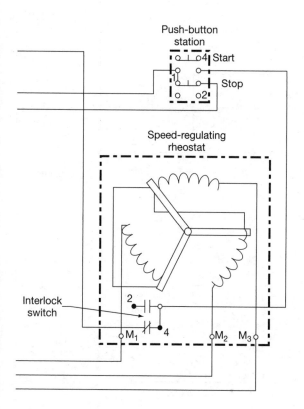

Push-button station

Start

Stop

Speed-regulating rheostat

Interlock switch

primary magnetic switch for a wound-rotor motor.

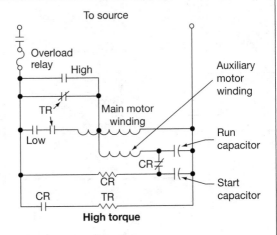

To source

Overload relay

High

TR

Low

Main motor winding

Auxiliary motor winding

Run capacitor

CR

Start capacitor

CR

CR   TR

High torque

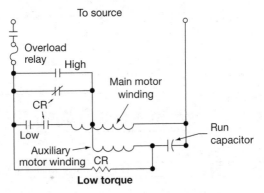

To source

Overload relay

High

CR

Low

Main motor winding

Auxiliary motor winding   CR

Run capacitor

Low torque

Starting and running circuits of low-torque and high-torque adjustable-speed capacitor motors with tapped main windings.

MOTOR CONTROL CIRCUITS (cont.)

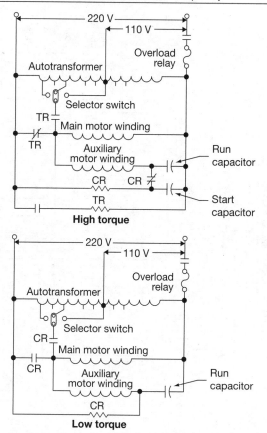

High torque

Low torque

Starting and running circuits of low-torque and high-torque adjustable-speed capacitor motors with manually operated transformer speed regulator.

MOTOR CONTROL CIRCUITS (cont.)

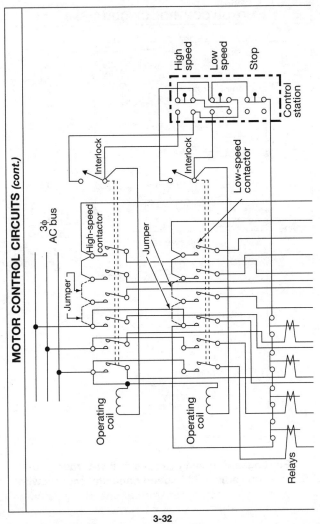

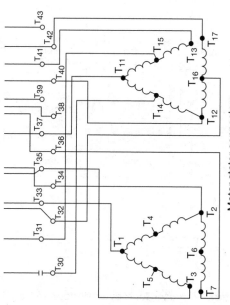

The control arrangement for an AC multi-speed motor.

Motor stator connectors

3-33

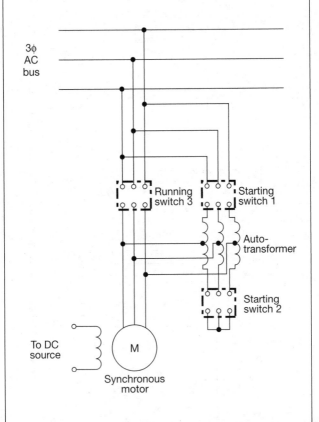

An autotransformer starter for a synchronous motor.

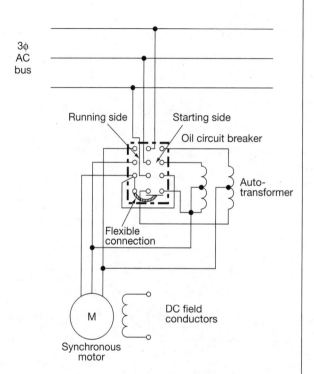

A reduced-voltage starter for a synchronous motor.

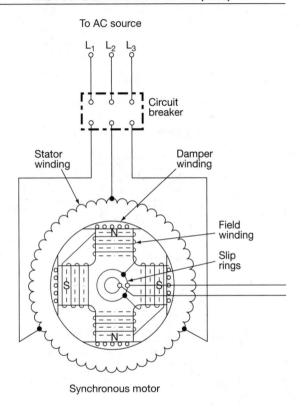

To AC source

L_1 L_2 L_3

Circuit breaker

Stator winding

Damper winding

Field winding

Slip rings

Synchronous motor

Connections of a synchronous motor and exciter with the

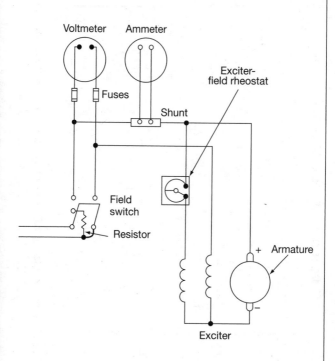

exciter-field rheostat, field switch, and exciter-field meters.

MOTOR CONTROL CIRCUITS (cont.)

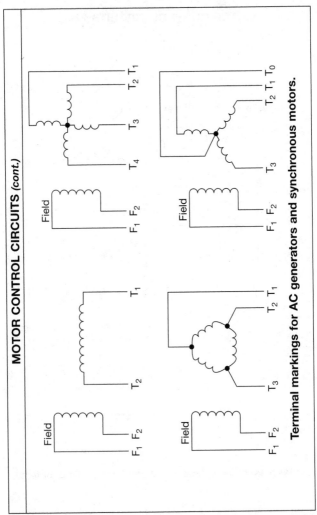

Terminal markings for AC generators and synchronous motors.

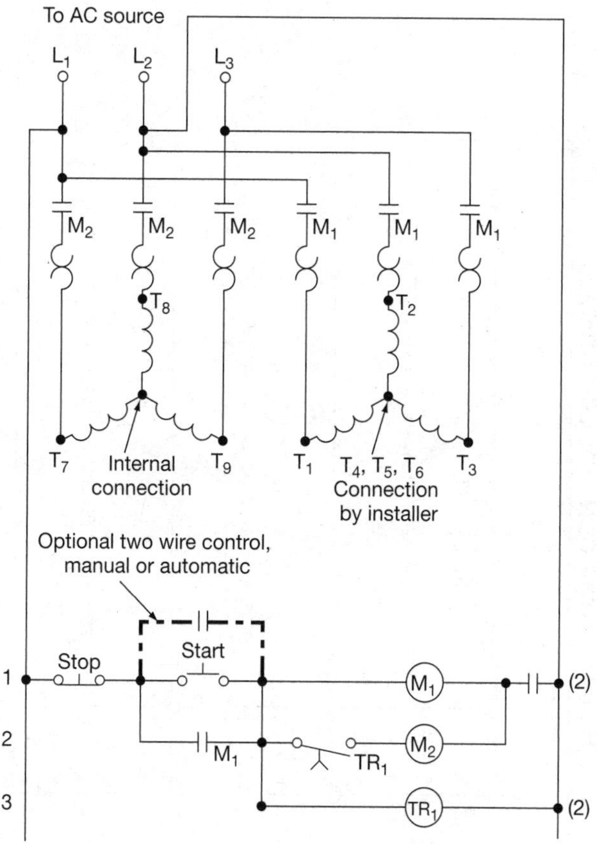

To AC source

L_1 L_2 L_3

M_2 M_2 M_2 M_1 M_1 M_1

T_8 T_2

T_7 Internal T_9 T_1 T_4, T_5, T_6 T_3
connection   Connection
by installer

Optional two wire control,
manual or automatic

	Start	

1   Stop   Start (M_1) (2)

2   M_1 TR₁ (M_2)

3   (TR_1) (2)

Part-winding reduced-voltage starting.

3-39

MOTOR CONTROL CIRCUITS (cont.)

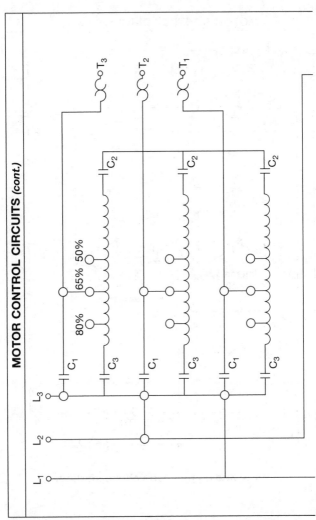

3-40

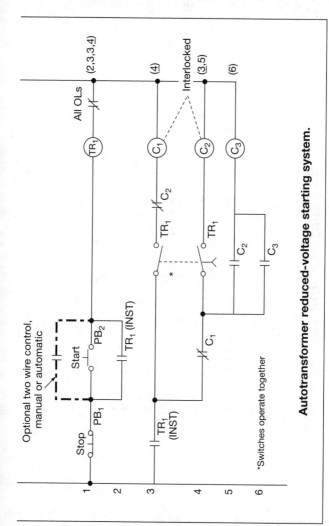

Autotransformer reduced-voltage starting system.

3-41

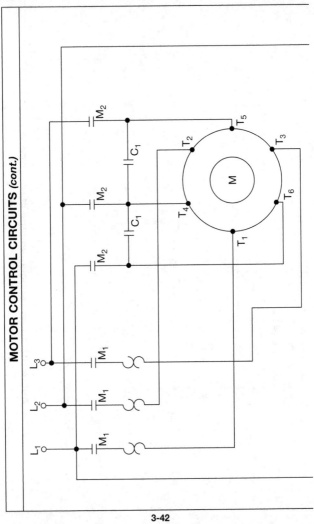

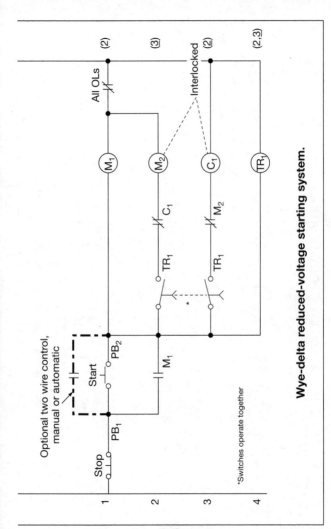

Wye-delta reduced-voltage starting system.

3-43

MOTOR CONTROL CIRCUITS *(cont.)*

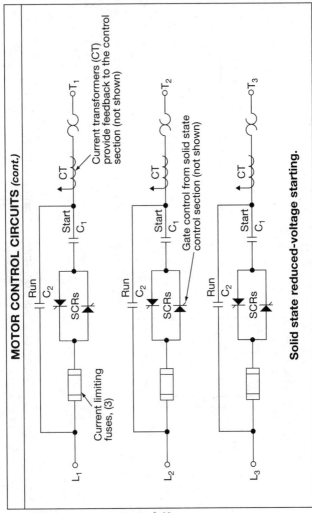

Current transformers (CT) provide feedback to the control section (not shown)

Gate control from solid state control section (not shown)

Current limiting fuses, (3)

Solid state reduced-voltage starting.

3-44

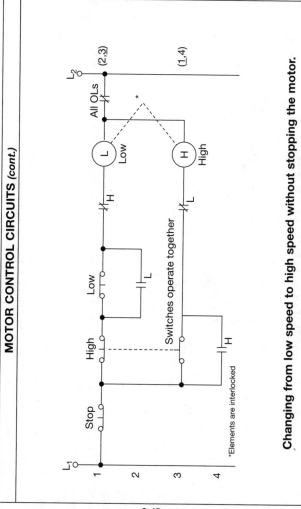

MOTOR CONTROL CIRCUITS (cont.)

Changing from low speed to high speed without stopping the motor.

*Elements are interlocked.

MOTOR CONTROL CIRCUITS (cont.)

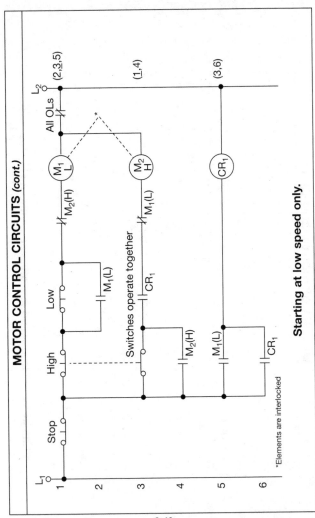

Starting at low speed only.

*Elements are interlocked

3-46

MOTOR CONTROL CIRCUITS (cont.)

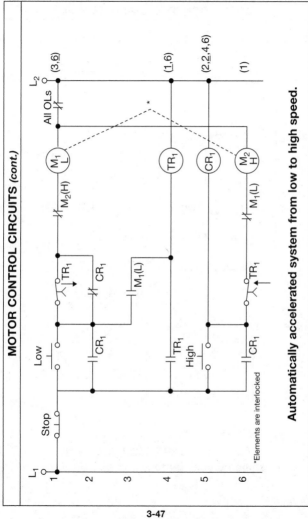

Automatically accelerated system from low to high speed.

*Elements are interlocked

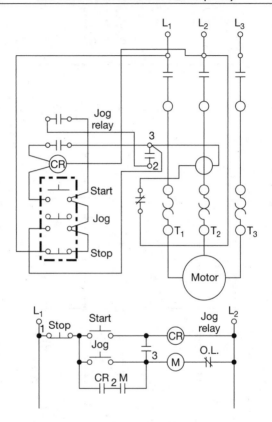

Separate "START-STOP-JOG" with standard push buttons and a JOG relay.

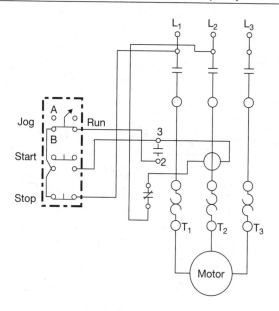

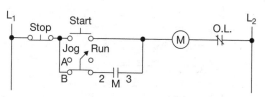

**Combined "START, JOG" and separate "STOP"
with selector switch. Jogging with a selector switch.**

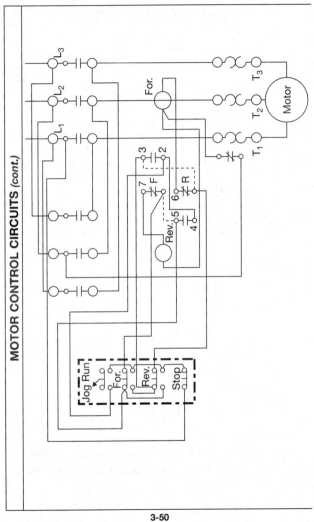

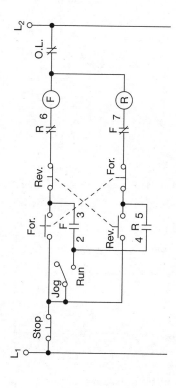

**Starting, Stopping and Jogging in either direction.
Jogging controlled through a Jogging selector switch.**

MOTOR CONTROL CIRCUITS (cont.)

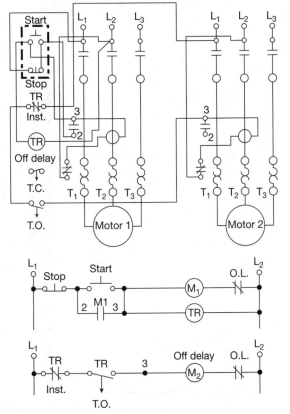

Sequence control of two motors — one to start and run for a short time after the other stops.

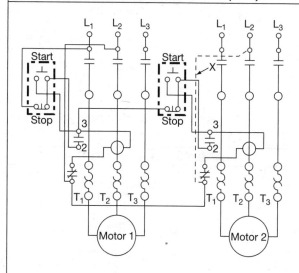

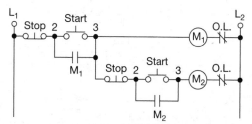

Control circuit is connected only to the lines of Motor 1.

Starters arranged for sequence control of a conveyor system.

DC SERIES-WOUND MOTOR

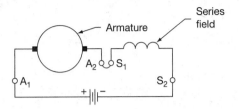

DC SHUNT-WOUND MOTOR

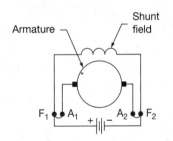

DC COMPOUND-WOUND MOTOR

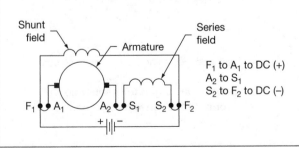

F_1 to A_1 to DC (+)
A_2 to S_1
S_2 to F_2 to DC (−)

DC MOTOR CONNECTIONS

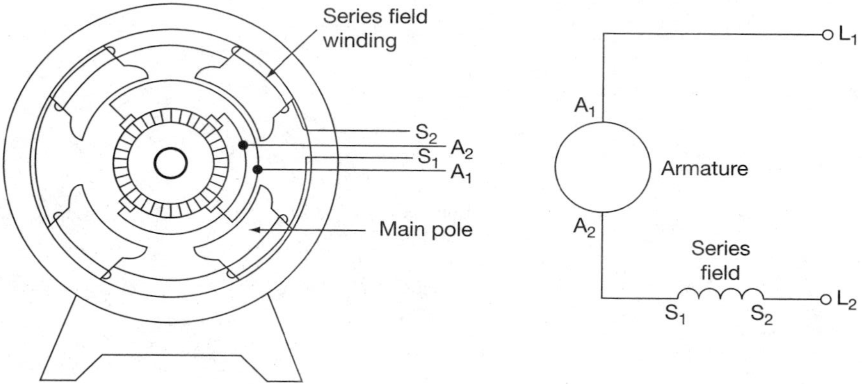
Series field
winding
S2
A2
S1
A1
Main pole
L1
A1
Armature
A2
Series
field
S1
S2
L2

Connections for a series-wound DC motor.

DC MOTOR CONNECTIONS (cont.)

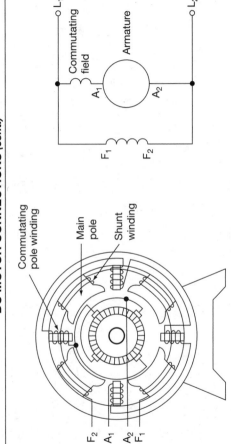

Shunt-wound DC motor with interpoles.

The interpoles (commutating poles) are in series with the armature so that their field strength will be proportional to the load on the motor.

DC MOTOR CONNECTIONS *(cont.)*

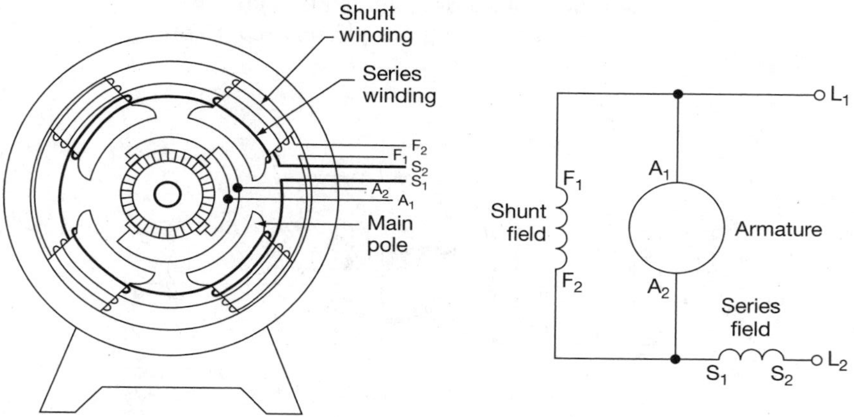

Shunt winding
Series winding
F2
F1
S2
S1
A2
A1
Main pole
L1
A1
F1
Shunt field
Armature
F2
A2
Series field
S1
S2
L2

Cumulatively compound-wound DC motor.

DC MOTOR CONNECTIONS *(cont.)*

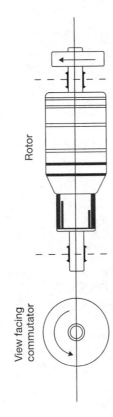

View facing commutator

Rotor

Standard direction of rotation is counterclockwise when facing commutator end of motor.

DC MOTOR CONNECTIONS (cont.)

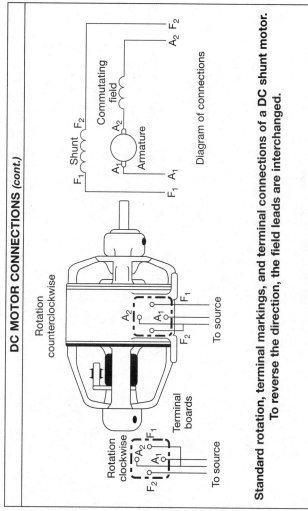

Standard rotation, terminal markings, and terminal connections of a DC shunt motor. To reverse the direction, the field leads are interchanged.

3-59

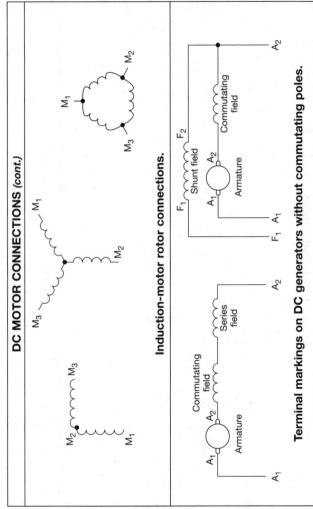

DC MOTOR CONNECTIONS (cont.)

M_1 M_2 M_3

Induction-motor rotor connections.

Commutating field

Armature

A_1 A_2 Series field A_2

A_1

F_2 Shunt field F_1

Commutating field A_2

Armature A_1

A_2

F_1

Terminal markings on DC generators without commutating poles.

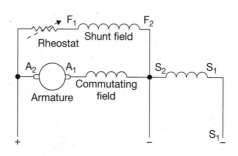

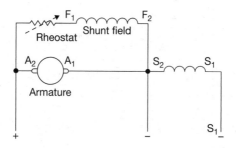

Terminal markings on DC compound generators.

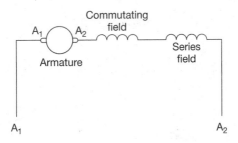

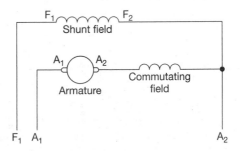

Terminal markings on nonreversing commutating-pole types of DC motors.

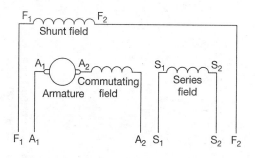

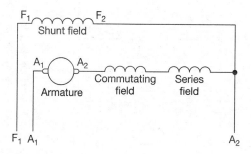

**Terminal markings on compound
commutating-pole types of DC motors.**

DC MOTOR CONTROL CIRCUITS

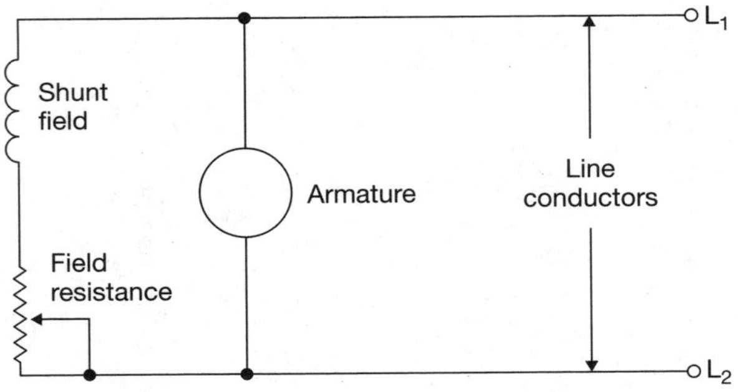

Speed control for a DC motor by connecting a variable resistance in series with the shunt field.

L₁

L₂

Line conductors

Armature

Shunt field

Field resistance

DC MOTOR CONTROL CIRCUITS (cont.)

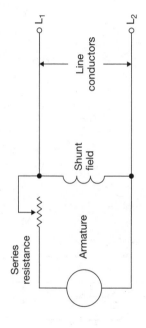

Speed control for a DC motor by connecting a variable resistance in series with the armature.

DC MOTOR CONTROL CIRCUITS *(cont.)*

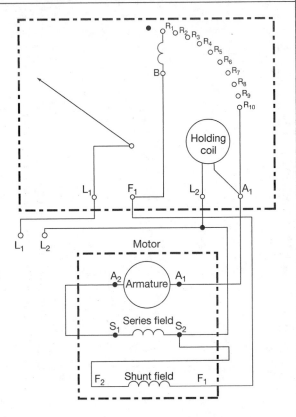

**Connections between a compound
DC motor and a faceplate starter.**

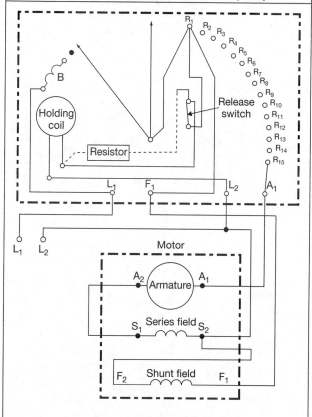

Connections between a compound DC motor and a speed-regulating faceplate starter.

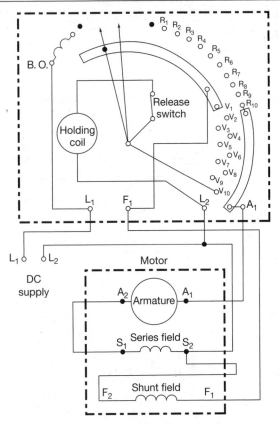

DC MOTOR CONTROL CIRCUITS (cont.)

B. O.

Release switch

Holding coil

R_1 R_2 R_3 R_4 R_5 R_6 R_7 R_8 R_9 R_{10}

V_1 V_2 V_3 V_4 V_5 V_6 V_7 V_8 V_9 V_{10}

L_1 F_1 L_2 A_1

L_1 L_2

DC supply

Motor

A_2 Armature A_1

S_1 Series field S_2

F_2 Shunt field F_1

Connections between a compound DC motor and a faceplate starter used also for speed control.

3-68

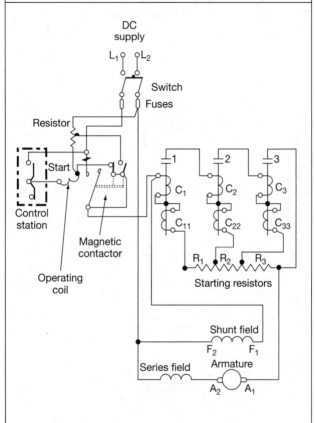

DC MOTOR CONTROL CIRCUITS *(cont.)*

DC supply

L_1 L_2

Switch

Fuses

Resistor

Start

Control station

Magnetic contactor

Operating coil

C_1 C_2 C_3

C_{11} C_{22} C_{33}

R_1 R_2 R_3

Starting resistors

Shunt field

F_2 F_1

Series field   Armature

A_2 A_1

Counter-emf DC motor starter.

DC MOTOR CONTROL CIRCUITS *(cont.)*

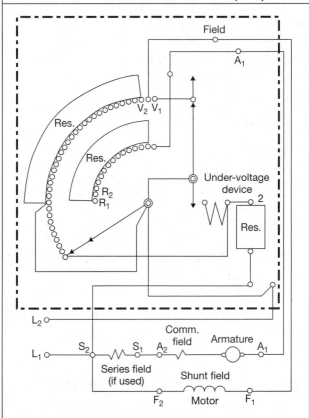

DC shunt motor speed-regulating rheostat for starting and speed control by field control.

DC MOTOR CONTROL CIRCUITS (cont.)

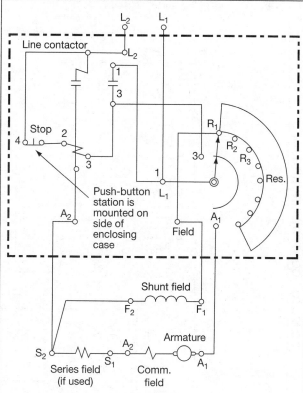

DC speed-regulating rheostat control for shunt or compound-wound motors with contactors and push-button station control.

DC MOTOR CONTROL CIRCUITS (cont.)

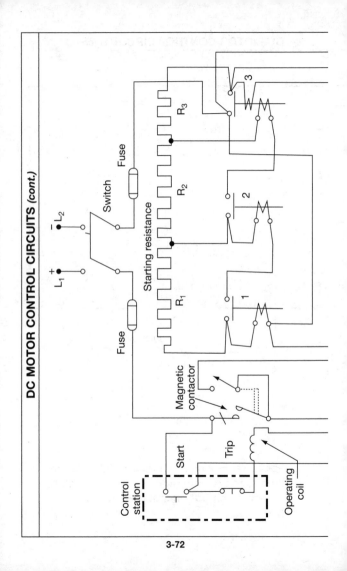

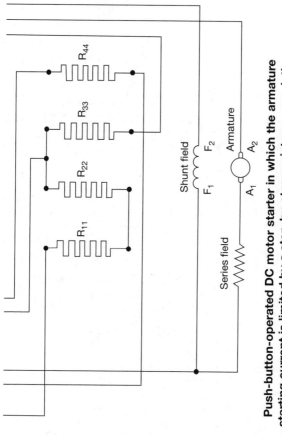

Push-button-operated DC motor starter in which the armature starting current is limited by a step-by-step resistance regulation.

DC MOTOR CONTROL CIRCUITS *(cont.)*

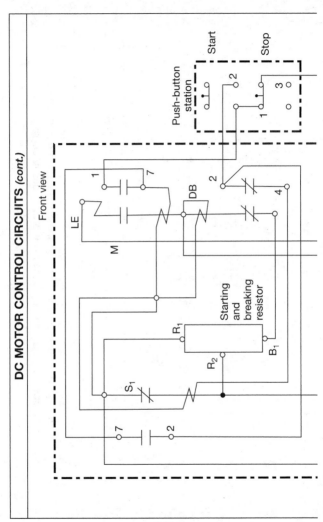

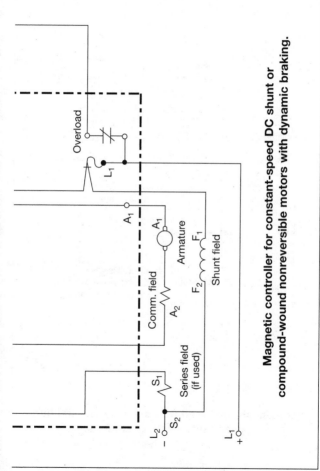

Magnetic controller for constant-speed DC shunt or compound-wound nonreversible motors with dynamic braking.

3-75

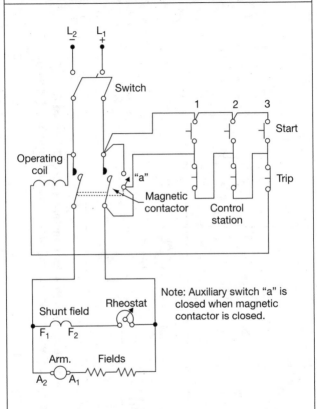

Magnetically operated DC motor starter with three push-button control stations.

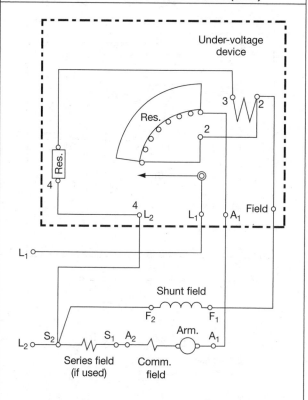

DC speed-regulating rheostat control for shunt or compound-wound motors without a contactor.

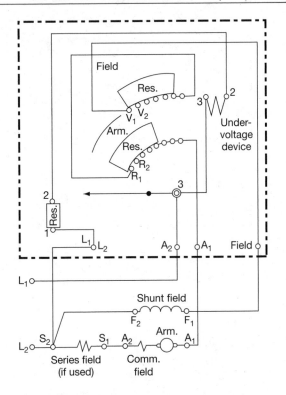

DC speed-regulating rheostat control for shunt or compound-wound motors; regulating duty — 50% speed reduction by armature control and 25% increase by field control.

CHAPTER 4
Design and Installation

It is important to understand the rules regarding motor application. The concerns are as follows:

1. Mechanical safety. This assures that motors are not a source of danger. For instance, we would not want to install open motors in areas where children could go. In the same manner, it is often necessary to put a clutch on a motor to avoid possible injury to a machine operator.

2. Mechanical stability and operations. Motors have mechanical stresses placed upon them. This results in vibration, which has the unfortunate side effect of loosening bolts and screws. This can also lead to other mechanical difficulties with the equipment that it operates.

3. Electrical safety. Motors should not become the source of an electrical shock or fault or cause problems to the electrical system on which they are installed.

4. Operational circuits. The circuits on which motors are installed must operate correctly. Motors place unusual demands on electrical circuits. They can cause large starting currents up to 8 times their normal full load current. They also put a lot of inductive reactance into electrical systems. And because of high current draw, they overheat electrical circuits more commonly than other types of loads.

4. Motor Controls. Suitable controllers are required for all motors. The simplest is the branch-circuit protective device for motors of $\frac{1}{8}$ horsepower or less. Another "controller" is a cord-and-plug connection for portable motors of $\frac{1}{3}$ horsepower or less. The concern with motor starters is the ability to close and open the contacts that connect the motor to the source. Unfortunately, it is not always possible to control the mechanical load applied to the motor. For this reason, overload relays are added to the motor starter. The current drawn by the motor is a reasonably accurate measure of the load on the motor, and thus of its heating.

Most overloads today use a thermally responsive element. The thermal element is connected mechanically to a normally closed contact. When excessive current flows long enough, the contact is tripped open. This contact is connected in series with the control coil of the starter. When the contact opens, the starter coil is deenergized. In turn, the starter power contacts disconnect the motor from the line. The overload heater element should be designed to have heat-storage characteristics similar to those of the motor. However, they should be just enough faster so that the relay will trip the normally closed relay contact before excessive heating occurs in the motor.

DESIGNING MOTOR CIRCUITS

For one motor:
1. Determine full-load current of motor(s).
2. Multiply full-load current x 1.25 to determine minimum conductor ampacity.
3. Determine wire size.
4. Determine conduit size.
5. Determine minimum fuse or circuit breaker size.
6. Determine overload rating.

For more than one motor:
1. Perform steps 1 through 6 as shown above for each motor.
2. Add full-load current of all motors, plus 25% of the full-load current of the largest motor to determine minimum conductor ampacity.
3. Determine wire size.
4. Determine conduit size.
5. Add the fuse or circuit breaker size of the largest motor, plus the full-load currents of all other motors to determine the maximum fuse or circuit breaker size for the feeder.

TIPS ON SELECTING MOTORS

Motors generally operate at their best in terms of power factor and efficiency when they are fully loaded. Motors generally operate at best power factor and efficiency when fully loaded.

TORQUE: Starting torque needed by load must be less than required starting torque of proposed motor. Motor torque must never fall below driven machine's torque needs in going from standstill to full speed.

Torque requirements of some loads may fluctuate between wide limits. Although average torque may be low, many torque peaks may be well above full-load torque. If load torque impulses are repeated frequently (air compressor), it's best to use a high-slip motor with a flywheel. But if load is generally steady at full load, you can use a more efficient low-slip motor. Only in this case any intermittent load peaks are taken directly by the motor and reflect back into power system. Also breakdown (maximum motor torque) must be higher than load-peak torque.

ENCLOSURES: Atmospheric conditions surrounding motor determine enclosure used. The more enclosed a motor, the more it costs and the hotter it tends to run. Totally-enclosed motors may require a larger frame size for a given HP than open or protected motors.

INSULATION: This, likewise, is determined by surrounding atmosphere and operating temperature. Ambient (room) temperature is generally assumed to be 40°C. Total temperature motor reaches directly influences insulation life. Each 10°C rise in max temp halves effective life of *Class A* and *B* insulations.

Motor temperature rise is maximum temperature (over ambient) measured with an external thermometer. "Hot-spot" allowance takes care of temperature difference between external reading and hottest spot within windings. Service-factor allows for continuous overload - 15% for general-purpose open, protected or drip-proof motors.

VARIABLE CYCLE: Where load varies according to some regular cycle, it would not be economical to select a motor that matches the peak load.

Instead (for AC induction motors where speed is not varied) calculate HP needed on the *root-means-square* (RMS) basis. *RMS* HP is equivalent continuos HP that would produce same heat in motor as cycle operation. Torque-speed relation of motor should still match that of load.

FACTS TO CONSIDER

REQUIREMENTS OF DRIVEN MACHINE:
1. HP needed
2. Torque range
3. Operating cycle-frequency of starts and stops
4. Speed
5. Operating position – horizontal, vertical or tilted
6. Direction of rotation
7. Endplay and thrust
8. Ambient (room) temperature
9. Surrounding conditions – water, gas, corrosion, dust, outdoor, etc.

ELECTRICAL SUPPLY:
1. Voltage of power system
2. Number of phases
3. Frequency
4. Limitations on starting current
5. Effect of demand, energy on power rates

SUMMARY OF MOTOR APPLICATIONS

DC and Single phase motors

Speed Regulation	Speed Control	Starting Torque	Pull-out Torque	Applications
Series				
Varies inversely as the load. Races on light loads and full voltage	Zero to maximum depending on control and load	High. Varies as square of the voltage. Limited by commutation, heating and line capacity	High. Limited by commutation, heating and line capacity	Where high starting torque is required and speed can be regulated. Traction, bridges, hoists, gates, car dumpers, car retarders, etc.
Shunt				
Drops 3 to 5% from no load to full load	Any desired range depending on motor design and type of system.	Good. With constant field, varies directly as voltage applied to armature	High. Limited by commutation, heating and line capacity	Where constant or adjustable speed is required and starting conditions are not severe. Fan, blowers, centrifugal pumps, conveyers, wood working machines, metal working machines, elevators
Compound				
Drops 7 to 20% from no load to full load depending on amount of compounding	Any desired range depending on motor design and type of control	Higher than for shunt, depending on amount of compounding	High. Limited by commutation, heating and line capacity	Where high starting torque combined with fairly constant speed is required. Plunger pumps, punch presses, shears, bending rolls, geared elevators, conveyors, hoists

SUMMARY OF MOTOR APPLICATIONS (*cont.*)

DC and Single phase motors

Speed Regulation	Speed Control	Starting Torque	Pull-out Torque	Applications
Capacitor				
Drops 5% for large to 10% for small sizes	None	150 to 350% of full load depending upon design and size	150% for large to 200% for small sizes	Constant speed service for any starting duty, and quiet operation, where polyphase current cannot be used
Commutator-Type				
Drops 5% for large to 10% for small sizes	Repulsion-induction, none. Brush shifting types 4 to 1 at full load	250% for large to 350% for small sizes	150% for large to 250% for small sizes	Constant speed service for any starting duty, where speed control is required and polyphase current cannot be used
Split-Phase				
Drops about 10% from no load to full load	None	75% for large to 175% for small sizes	150% for large to 200% for small sizes	Constant speed service where starting is easy, Small fans, centrifugal pumps, and light running machines, where polyphase current is not available

SUMMARY OF MOTOR APPLICATIONS (*cont.*)

2 and 3 phase motors

Speed Regulation	Speed Control	Starting Torque	Pull-out Torque	Applications
General-purpose squirrel-cage (Class B)				
Drops about 3% for large to 5% for small sizes	None, except multi-speed types designed for 2 to 4 fixed speeds	200% of full load for 2-pole to 105% for 16 pole designs	200% of full load	Constant-speed service where starting torque is not excessive fans, blowers, rotary compressors, centrifugal pumps
High torque squirrel-cage (Class C)				
Drops about 3% for large to 6% for small sizes	None, except multi-speed types designed for 2 to 4 fixed speeds	250% of full load for high speed to 200% for low speed designs	200% of full load	Constant-speed service where fairly high starting torque is required at infrequent intervals with starting current of about 400% of full load. Reciprocating pumps and compressors, crushers, etc.
High slip squirrel-cage (Class D)				
Drops about 10 to 15% from no load to full load	None, except multi-speed types designed for 2 to 4 fixed speeds	225 to 300% of full load, depending on speed with rotor resistance	200% will usually not stall until loaded to max. torque, which occurs at standstill	Constant-speed service and high starting torque, if starting is not too frequent, and for taking high peak loads with or without flywheels. Punch presses, shears and elevators, etc.

SUMMARY OF MOTOR APPLICATIONS (*cont.*)

2 and 3 phase motors

Speed Regulation	Speed Control	Starting Torque	Pull-out Torque	Applications
Low torque squirrel-cage (Class F)				
Drops about 3% for large to 5% for small sizes	None, except multi-speed types designed for 2 to 4 fixed speeds	50% of full load for high speed to 90% for low speed designs	150 to 170% of full load	Constant speed service where starting duty is light. Fans blowers, centrifugal pumps and similar loads
Wound-rotor				
With rotor rings short circuited, drops about 3% for large to 5% for small sizes	Speed can be reduced to 50% by rotor resistance to obtain stable operation. Speed varies inversely as load	Up to 300% depending on external resistance in rotor circuit and how distributed	200% when rotor slip rings are short circuited	Where high starting torque with low starting current or where limited speed control is required. Fans, centrifugal and plunger pumps, compressors, conveyers, hoists, cranes
Synchronous				
Constant	None, except special motors designed for 2 fixed speeds	40% for slow to 160% for medium speed 80% pf designs. Special designs develop higher torques	Unity-of motors 170%; 80% pf motors 225%. Special designs up to 300%	For constant-speed service, direct connection to slow speed machines and where power factor correction is required.

THREE-PHASE MOTOR REQUIREMENTS

Motor HP	Amperes at Full Load		Size of Conduit In Inches		Minimum Wire Size Rubber – AWG	
	230 Volts	460 Volts	230 Volts	460 Volts	230 Volts	460 Volts
1	3.3	1.7	½	½	14	14
1½	4.7	2.4	½	½	14	14
2	6	3	½	½	14	14
3	9	4.5	½	½	14	14
5	15	7.5	½	½	12	14
7½	22	11	¾	½	8	14
10	27	14	¾	½	8	12
15	38	19	1¼	¾	6	10
20	52	26	1¼	¾	4	8
25	64	32	1¼	1¼	3	6
30	77	39	1½	1¼	1	6
40	101	51	2	1¼	2/0	4
50	125	63	2	1¼	3/0	3
60	149	75	2½	1½	200 kcmil	1
75	180	90	2½	2	4/0	1/0
100	245	123	3	2	500 kcmil	3/0
125	310	155	3½	2½	750 kcmil	4/0
150	360	180	4	2½	1000 kcmil	300 kcmil
200	480	240	—	3	—	500 kcmil

If voltage drop exceeds limits, the wire size should be adjusted.   kcmil = 1000 circular mils

DIRECT CURRENT MOTOR REQUIREMENTS

Motor HP	Amperes at Full Load		Size of Conduit In Inches		Minimum Wire Size Rubber – AWG	
	115 Volts	230 Volts	115 Volts	230 Volts	115 Volts	230 Volts
1	8.4	4.2	½	½	14	14
1½	12.5	6.3	½	½	12	14
2	16.1	8.3	¾	½	10	14
3	23	12.3	¾	½	8	12
5	40	19.8	1	¾	6	10
7½	58	28.7	1¼	1	3	6
10	75	38	1½	1	1	6
15	112	56	2	1¼	2/0	4
20	140	74	2	1½	3/0	1
25	184	92	2½	2	300 kcmil	1/0
30	220	110	3	2	400 kcmil	2/0
40	292	146	3½	2½	700 kcmil	4/0
50	360	180	4	2½	1000 kcmil	300 kcmil
60	—	215	—	3	—	400 kcmil
75	—	268	—	3½	—	600 kcmil
100	—	355	—	4	—	1000 kcmil

kcmil = 1000 circular mils

4-9

1φ 115 V MOTORS AND CIRCUITS – 120 V SYSTEM

Size of motor HP	Amp	Motor overload protection Low-peak or Fusetron® — Motor less than 40°C or greater than 1.15 SF (Max. fuse 125%)	All other motors (Max. fuse 115%)	Switch 115% minimum or HP rated or fuse holder size	Minimum size of starter	Controller termination temperature rating 60°C TW	60°C THW	75°C TW	75°C THW	Wire size (AWG or kcmil)	Conduit (inches)
1/6	4.4	5	5	30	00	•	•	•	•	14	1/2
1/4	5.8	7	6 1/4	30	00	•	•	•	•	14	1/2
1/3	7.2	9	8	30	00	•	•	•	•	14	1/2
1/2	9.8	12	10	30	00	•	•	•	•	14	1/2
3/4	13.8	15	15	30	00	•	•	•	•	12	1/2
1	16	20	17 1/2	30	00	•	•	•	•	12	1/2
1 1/2	20	25	20	30	01	•	•	•	•	10	1/2
2	24	30	25	30	01	•	•	•	•	10	1/2

4-10

1∅ 230 V MOTORS AND CIRCUITS – 240 V SYSTEM

| Size of motor | | Motor overload protection Low-peak or Fusetron® | | Switch 115% minimum or HP rated or fuse holder size | Minimum size of starter | Controller termination temperature rating | | | | Minimum size of copper wire and trade conduit | |
| | | | | | | 60°C | | 75°C | | | |
HP	Amp	Motor less than 40°C or greater than 1.15 SF (Max. fuse 125%)	All other motors (Max. fuse 115%)			TW THW		TW THW		Wire size (AWG or kcmil)	Conduit (inches)
1/6	2.2	2 1/2	2 1/2	30	00	•		•		14	1/2
1/4	2.9	3 1/2	3 2/10	30	00	•		•		14	1/2
1/3	3.6	4 1/2	4	30	00	•		•		14	1/2
1/2	4.9	5 6/10	5 6/10	30	00	•		•		14	1/2
3/4	6.9	8	7 1/2	30	00	•		•		14	1/2
1	8	10	9	30	00	•		•		14	1/2
1 1/2	10	12	10	30	00	•		•		14	1/2
2	12	15	12	30	0	•		•		14	1/2
3	17	20	17 1/2	30	1	•		•		12	1/2

4-11

1φ 230 V MOTORS AND CIRCUITS – 240 V SYSTEM (cont.)

Size of motor HP	Amp	Motor overload protection Low-peak or Fusetron® — Motor less than 40°C or greater than 1.15 SF (Max. fuse 125%)	All other motors (Max. fuse 115%)	Switch 115% minimum or HP rated or fuse holder size	Minimum size of starter	Controller termination temperature rating 60°C TW	60°C THW	75°C TW	75°C THW	Wire size (AWG or kcmil)	Conduit (inches)
5	28	35	30*	60	2	•	•			8	3/4
								•		8	1/2
									•	10	1/2
7 1/2	40	50	45	60	2	•	•			6	3/4
								•	•	8	3/4
10	50	60	50	60	3	•	•			4	1
									•	6	3/4

* Fuse reducers required.

4-12

3φ 230 V MOTORS AND CIRCUITS – 240 V SYSTEM

Size of motor HP	Amp	Motor overload protection Low-peak or Fusetron®		Switch 115% minimum or HP rated or fuse holder size	Minimum size of starter	Controller termination temperature rating				Minimum size of copper wire and trade conduit	
		Motor less than 40°C or greater than 1.15 SF (Max. fuse 125%)	All other motors (Max. fuse 115%)			60°C		75°C		Wire size (AWG or kcmil)	Conduit (inches)
						TW	THW	TW	THW		
1/2	2	2 1/2	2 1/4	30	00	•	•	•	•	14	1/2
3/4	2.8	3 1/2	3 2/10	30	00	•	•	•	•	14	1/2
1	3.6	4 1/2	4	30	00	•	•	•	•	14	1/2
1 1/2	5.2	6 1/4	5 6/10	30	00	•	•	•	•	14	1/2
2	6.8	8	7 1/2	30	0	•	•	•	•	14	1/2
3	9.6	12	10	30	0	•	•	•	•	14	1/2
5	15.2	17 1/2	17 1/2	30	1	•	•	•	•	14	1/2
7 1/2	22	25	25	30	1	•	•	•	•	10	1/2
10	28	35	30*	60	2	•				8	3/4
									•	10	1/2
15	42	50	45	60	2	•				6	1
									•	6	3/4

* Fuse reducers required.

4-13

3φ 230 V MOTORS AND CIRCUITS – 240 V SYSTEM (cont.)

Size of motor HP	Amp	Motor overload protection Low-peak or Fusetron® — Motor less than 40°C or greater than 1.15 SF (Max. fuse 125%)	All other motors (Max. fuse 115%)	Switch 115% minimum or HP rated or fuse holder size	Minimum size of starter	60°C TW	60°C THW	75°C TW	75°C THW	Wire size (AWG or kcmil)	Conduit (inches)
20	54	60*	60*	100	3	•	•			4	1
								•	•	3	1 1/2
25	68	80	75	100	3	•	•			3	1
										4	1
30	80	100	90	100	3	•	•			1	1 1/4
								•	•	3	1 1/4
40	104	125	110	200	4	•	•			2/0	1 1/2
								•	•	1	1 1/4
50	130	150	150	200	4	•	•			3/0	2
								•	•	2/0	1 1/2

Controller termination temperature rating columns: 60°C (TW, THW) and 75°C (TW, THW)

Minimum size of copper wire and trade conduit

* Fuse reducers required.

4-14

3φ 230 V MOTORS AND CIRCUITS – 240 V SYSTEM (cont.)

Size of motor HP	Amp	Motor overload protection Low-peak or Fusetron®: Motor less than 40°C or greater than 1.15 SF (Max. fuse 125%)	All other motors (Max. fuse 115%)	Switch 115% minimum or HP rated or fuse holder size	Minimum size of starter	Controller termination temperature rating 60°C TW	60°C THW	75°C TW	75°C THW	Wire size (AWG or kcmil)	Conduit (inches)
75	192	225	200*	400	5	•	•		•	300	2 1/2
									•	250	2 1/2
100	248	300	250	400	5	•	•		•	500	3
									•	350	2 1/2
150	360	450	400*	600	6		•		•	300-2/φ**	2–2 1/2**
									•	4/0-2/φ**	2-2**

* Fuse reducers required.
** Two runs of conduit and two sets of multiple conductors are required.

4-15

3φ 460 V MOTORS AND CIRCUITS – 480 V SYSTEM

Size of motor HP	Amp	Motor overload protection Low-peak or Fusetron® — Motor less than 40°C or greater than 1.15 SF (Max. fuse 125%)	All other motors (Max. fuse 115%)	Switch 115% minimum or HP rated or fuse holder size	Minimum size of starter	Controller termination temperature rating 60°C TW	60°C THW	75°C TW	75°C THW	Wire size (AWG or kcmil)	Conduit (inches)
1/2	1	1 1/4	1 1/8	30	00	•	•	•	•	14	1/2
3/4	1.4	1 6/10	1 6/10	30	00	•	•	•	•	14	1/2
1	1.8	2 1/4	2	30	00	•	•	•	•	14	1/2
1 1/2	2.6	3 2/10	2 6/10	30	00	•	•	•	•	14	1/2
2	3.4	4	3 1/2	30	00	•	•	•	•	14	1/2
3	4.8	5 6/10	5	30	0	•	•	•	•	14	1/2
5	7.6	9	8	30	0	•	•	•	•	14	1/2
7 1/2	11	12	12	30	1	•	•	•	•	14	1/2
10	14	17 1/2	15	30	1	•	•	•	•	14	1/2
15	21	25	20	30	2	•	•	•	•	10	1/2
20	27	30*	30*	60	2				•	8 / 10	3/4 / 1/2

* Fuse reducers required.

4-16

3φ 460 V MOTORS AND CIRCUITS – 480 V SYSTEM (cont.)

Size of motor HP	Amp	Motor overload protection Low-peak or Fusetron® Motor less than 40°C or greater than 1.15 SF (Max. fuse 125%)	All other motors (Max. fuse 115%)	Switch 115% minimum or HP rated or fuse holder size	Minimum size of starter	Controller termination temperature rating 60°C TW	60°C THW	75°C TW	75°C THW	Wire size (AWG or kcmil)	Conduit (inches)
25	34	40	35	60	2	•	•	•	•	6	1
										8	3/4
30	40	50	45	60	3	•	•	•	•	6	1
										8	3/4
40	52	60*	60*	100	3	•	•	•	•	4	1
										6	1
50	65	80	70	100	3	•	•	•	•	3	1 1/4
										4	1
60	77	90	80	100	4	•	•	•	•	1	1 1/4
										3	1 1/4
75	96	110	110	200	4	•	•	•	•	1/0	1 1/2
										1	1 1/4

* Fuse reducers required.

4-17

3φ 460 V MOTORS AND CIRCUITS – 480 V SYSTEM (cont.)

Size of motor HP	Amp	Motor overload protection Low-peak or Fusetron® Motor less than 40°C or greater than 1.15 SF (Max. fuse 125%)	All other motors (Max. fuse 115%)	Switch 115% minimum or HP rated or fuse holder size	Minimum size of starter	60°C TW	60°C THW	75°C TW	75°C THW	Wire size (AWG or kcmil)	Conduit (inches)
100	124	150	125	200	4	•	•			3/0	2
								•	•	2/0	1 1/2
125	156	175	175	200	5	•	•			4/0	2
								•	•	3/0	2
150	180	225	200*	400	5	•	•			300	2 1/2
								•	•	4/0	2
200	240	300	250	400	5	•	•			500	3
								•	•	350	2 1/2
250	302	350	325	400	6	•	•			4/0-2/φ**	2-2**
								•	•	3/0-2/φ**	2-2**
300	361	450	400*	600	6	•	•			300-2/φ**	2-1 1/2**
								•	•	4/0-2/φ**	2-2**

* Fuse reducers required.
**Two runs of conduit and two sets of multiple conductors are required.

DC MOTORS AND CIRCUITS

Size of motor HP	Amp	Motor overload protection Low-peak or Fusetron® — Motor less than 40°C or greater than 1.15 SF (Max. fuse 125%)	All other motors (Max. fuse 115%)	Switch 115% minimum or HP rated or fuse holder size	Minimum size of starter	Controller termination temperature rating 60°C TW	60°C THW	75°C TW	75°C THW	Minimum size of copper wire and trade conduit — Wire size (AWG or kcmil)	Conduit (inches)
90 V											
1/4	4.0	5	4 1/2	30	0	•	•	•	•	14	1/2
1/3	5.2	6 1/4	5 6/10	30	0	•	•	•	•	14	1/2
1/2	6.8	8	7.5	30	0	•	•	•	•	14	1/2
3/4	9.6	12	10	30	0	•	•	•	•	14	1/2
1	12.2	15	12	30	0	•	•	•	•	14	1/2
120 V											
1/4	3.1	3 1/2	3 1/2	30	0	•	•	•	•	14	1/2
1/3	4.1	5	4 1/2	30	0	•	•	•	•	14	1/2
1/2	5.4	6 1/4	6	30	0	•	•	•	•	14	1/2
3/4	7.6	9	8	30	0	•	•	•	•	14	1/2
1	9.5	10	10	30	0	•	•	•	•	14	1/2

* Fuse reducers required.

DC MOTORS AND CIRCUITS (cont.)

Size of motor HP	Amp	Motor overload protection Low-peak or Fusetron® — Motor less than 40°C or greater than 1.15 SF (Max. fuse 125%)	All other motors (Max. fuse 115%)	Switch 115% minimum or HP rated or fuse holder size	Minimum size of starter	60°C TW	60°C THW	75°C TW	75°C THW	Wire size (AWG or kcmil)	Conduit (inches)
120 V (cont.)											
1 1/2	13.2	15	15	30	1	•	•	•	•	14	1/2
2	17	20	17 1/2	30	1	•	•	•	•	12	1/2
5	40	50	45	60	2	•	•			6	3/4
								•	•	8	3/4
10	76	90	80	100	3	•	•			2	1
								•	•	3	1
180 V											
1/4	2	2 1/2	2 1/4	30	0	•	•	•	•	14	1/2
1/3	2.6	3 2/10	2 8/10	30	0	•	•	•	•	14	1/2
1/2	3.4	4	3 1/2	30	0	•	•	•	•	14	1/2
3/4	4.8	6	5	30	0	•	•	•	•	14	1/2

* Fuse reducers required.

DC MOTORS AND CIRCUITS (cont.)

Size of motor		Motor overload protection Low-peak or Fusetron®		Switch 115% minimum or HP rated or fuse holder size	Minimum size of starter	Controller termination temperature rating				Minimum size of copper wire and trade conduit	
		Motor less than 40°C or greater than 1.15 SF (Max. fuse 125%)	All other motors (Max. fuse 115%)			60°C		75°C		Wire size (AWG or kcmil)	Conduit (inches)
HP	Amp					TW	THW	TW	THW		
180 V (cont.)											
1	6.1	7 1/2	7	30	0	•	•	•	•	14	1/2
1 1/2	8.3	10	9	30	1	•	•	•	•	14	1/2
2	10.8	12	12	30	1	•	•	•	•	14	1/2
3	16	20	17 1/2	30	1	•	•	•	•	12	1/2
5	27	30*	30*	60	1			•	•	8	1/2
						•				8	3/4

* Fuse reducers required.

4-21

CONTROL RATINGS

Size	Load (V)	Max HP Normal duty 1φ	Max HP Normal duty 3φ	Max HP Plugging & jogging duty 1φ	Max HP Plugging & jogging duty 3φ	Cont. amps	Service limit amps	Tungsten & ballast type lamp amps 480 V max.	Resistance heating (kW) 1φ	Resistance heating (kW) 3φ	Transformer switching 20 times 1φ	Transformer switching 20 times 3φ	Transformer switching 20-40 times 1φ	Transformer switching 20-40 times 3φ	Capacitor kVA switching rating 3φ kVAR
00	115	1/2	—	—	—	9	11	—	1.15	2.0	—	—	—	—	—
	200	1	1 1/2	—	—	9	11	—	2.0	3.46	—	—	—	—	—
	230	1	1 1/2	—	—	9	11	—	2.3	4.0	—	—	—	—	—
	380	—	1 1/2	—	—	9	11	—	—	6.5	—	—	—	—	—
	460	—	2	—	—	9	11	—	4.6	8.0	—	—	—	—	—
	575	—	2	—	—	9	11	—	5.8	10.0	—	—	—	—	—
0	115	1	—	1/2	—	18	21	20	2.3	4.0	—	—	—	—	—
	200	2	3	1	1 1/2	18	21	20	4.0	6.92	0.6	1.8	0.3	0.9	—
	230	2	3	1	1 1/2	18	21	20	4.6	8.0	1.2	2.1	0.6	1.0	—
	380	—	5	—	1 1/2	18	21	20	—	13.1	—	—	—	—	—
	460	—	5	—	2	18	21	20	9.2	15.9	2.4	4.2	1.2	2.1	—
	575	—	5	—	2	18	21	—	11.5	19.9	3.0	5.2	1.5	2.6	—
1	115	2	—	1	—	27	32	30	3.5	6.0	—	—	—	—	—
	200	3	7 1/2	2	3	27	32	30	6	10.4	1.2	3.6	0.6	1.8	—
	230	3	7 1/2	2	3	27	32	30	6.9	11.9	2.4	4.3	1.2	2.1	—
	380	—	10	—	5	27	32	30	—	19.7	—	—	—	—	—
	460	—	10	—	5	27	32	30	13.8	23.9	4.9	8.5	2.5	4.3	—
	575	—	10	—	5	27	32	—	17.3	29.8	6.2	11.0	3.1	5.3	—

CONTROL RATINGS (cont.)

Size	Load (V)	Maximum HP — Normal duty 1φ	Normal duty 3φ	Plugging & jogging duty 1φ	Plugging & jogging duty 3φ	Cont. amps	Service limit amps	Tungsten & ballast type lamp amps 480 V max.	Resistance heating (kW) 1φ	Resistance heating (kW) 3φ	Transformer switching 50—60 Hz kVA rating inrush peak time Continuous amps — 20 times 1φ	20 times 3φ	20–40 times 1φ	20–40 times 3φ	Capacitor kVA switching ratings 3φ kVAR
1P	115	3	—	1 1/2	—	35	42	45	5.8	—	—	—	—	—	—
	230	5	—	3	—	35	42	45	11.5	—	—	—	—	—	—
1 3/4	115		—		—	40	40	45	5.8	9.9	1.6	—	0.8	—	—
	200		10		5	40	40	45	10	17.3	—	4.9	—	2.4	—
	230		10		7 1/2	40	40	45	11.5	19.9	3.2	5.75	1.6	2.8	—
	380		15		7 1/2	40	40	45	—	32.9	—	—	—	—	—
	460		15		7 1/2	40	40	45	23	39.8	6.6	11.2	3.3	5.7	—
	575		15		7 1/2	40	40	—	28.8	49.7	8.1	14.5	4.1	7.1	8
2	115	3	—	2	—	45	52	60	8.1	13.9	2.1	—	1.0	—	—
	200		10		7 1/2	45	52	60	14	24.2	—	6.3	—	3.1	—
	230	7 1/2	15	5	10	45	52	60	16.1	27.8	4.1	7.2	2.1	3.6	—
	380		25		15	45	52	60	—	46.0	—	—	—	—	—
	460		25		15	45	52	60	32.2	55.7	8.3	14	4.2	7.2	16
	575		25		15	45	52	—	40.3	69.6	10.0	18	5.2	8.9	20
2 1/2	115	5	—		—	60	65	75	10.4	17.9	3.1	—	1.5	—	—
	200		15		10	60	65	75	18	31.1	—	9.1	—	4.6	—
	230	10	20		15	60	65	75	20.7	35.8	6.1	10.6	3.1	5.3	17.5
	380		30		20	60	65	75	—	59.2	—	—	—	—	—
	460		30		20	60	65	75	41.4	71.6	12	21	6.1	10.6	34.5
	575		30		20	60	65	—	51.8	89.5	15	26.5	7.6	13.4	43.5

CONTROL RATINGS (cont.)

Size	Load (V)	Maximum HP Normal duty 1φ	Normal duty 3φ	Plugging & jogging duty 1φ	Plugging & jogging duty 3φ	Cont. amps	Service limit amps	Tungsten & ballast type lamp amps 480 V max.	Resistance heating (kW) 1φ	Resistance heating 3φ	Transformer switching 20 times 1φ	20 times 3φ	20-40 times 1φ	20-40 times 3φ	Capacitor kVA switching ratings 3φ kVAR
3	115	7 1/2	—	—	—	90	104	100	14.4	24.8	4.1	—	2.0	—	—
	200	—	25	—	15	90	104	100	25	43.3	—	12	—	6.1	—
	230	15	30	—	20	90	104	100	28.8	50.0	8.1	14	4.1	7.0	27
	380	—	50	—	30	90	104	100	—	82.2	—	—	—	—	—
	460	—	50	—	30	90	104	100	57.5	99.4	16	28	8.1	14	53
	575	—	50	—	30	90	104	—	71.9	124	20	35	10	18	67
3 1/2	115	—	—	—	—	115	125	150	18.4	31.8	—	—	—	—	—
	200	—	30	—	20	115	125	150	32	55.4	—	16	—	8	—
	230	—	60	—	25	115	125	150	36.8	63.7	11	18.5	5.4	9.5	33.5
	380	—	60	—	30	115	125	150	—	105	—	—	—	—	—
	460	—	75	—	40	115	125	150	73.6	127	21.5	37.5	11.0	18.5	66.5
	575	—	75	—	40	115	125	—	92	159	37	47	13.5	23.5	83.5
4	200	—	40	—	25	135	156	200	39	67.5	—	20	—	10	—
	230	—	50	—	30	135	156	200	44.9	77.6	—	23	—	12	40
	380	—	75	—	50	135	156	200	—	128	14	—	6.8	—	—
	460	—	100	—	60	135	156	200	89.7	155	27	47	14	23	80
	575	—	100	—	60	135	156	—	112	194	34	59	17	29	100

CONTROL RATINGS (cont.)

Size	Load (V)	Maximum HP Normal duty 1φ	Maximum HP Normal duty 3φ	Maximum HP Plugging & jogging duty 1φ	Maximum HP Plugging & jogging duty 3φ	Cont. amps 3φ	Service limit amps	Tungsten & ballast type lamp amps 480 V max.	Resistance heating (kW) 1φ	Resistance heating (kW) 3φ	Transformer switching 20 times 1φ	Transformer switching 20 times 3φ	Transformer switching 20-40 times 1φ	Transformer switching 20-40 times 3φ	Capacitor kVA switching ratings 3φ kVAR
	200	—	50	—	30	210	225	250	53	91.7	—	30.5	—	15	—
	230	—	75	—	40	210	225	250	60.9	105	20.5	35	10.4	18	60
4 1/2	380	—	100	—	75	210	225	250	—	174	—	—	—	—	—
	460	—	150	—	100	210	225	250	122	211	40.5	70.5	20.5	35	120
	575	—	150	—	100	210	225	—	152	264	51	88	25.5	44	150

STANDARD MOTOR SIZES

Classification	Size (HP)
Milli	1, 1.5, 2, 3, 5, 7.5, 10, 15, 25, 35
Fractional	1/20, 1/12, 1/8, 1/6, 1/4, 1/3, 1/2, 3/4
Full	1, 1-1/2, 2, 3, 5, 7-1/2, 10, 15, 20, 25, 30, 40, 50, 60, 75, 100, 125, 150, 200, 250, 300
Full—Special Order	350, 400, 450, 500, 600, 700, 800, 900, 1000, 1250, 1500, 1750, 2000, 2250, 2500, 3000, 3500, 4000, 4500, 5000, 5500, 6000, 7000, 8000, 9000, 10,000, 12,000, 13,000, 14,000, 15,000, 16,000, 17,000, 18,000, 19,000, 20,000, 22,500, 30,000, 32,500, 35,000, 37,500, 40,000, 45,000, 50,000

STARTING METHODS: SQUIRREL-CAGE INDUCTION MOTORS

Starter Type	% Full-Voltage Value		
	Voltage at Motor	Line Current	Motor Output Torque
Full Voltage	100	100	100
Autotransformer:			
80 pc tap	80	64*	64
65 pc tap	65	42*	42
50 pc tap	50	25*	25
Primary-reactor:			
80 pc tap	80	80	64
65 pc tap	65	65	42
50 pc tap	50	50	25
Primary-resistor:			
Typical rating	80	80	64
Part-winding:			
Low-speed motors (½ -½)	100	50	50
High-speed motors (½ -½)	100	70	50
High-speed motors (⅔ - ⅓)	100	65	42
Wye Start-Delta Run (⅓ - ⅓)	100	33	33

*Autotransformer magnetizing current not included.
Magnetizing current usually less than 25 percent motor full-load current

SETTING BRANCH CIRCUIT PROTECTIVE DEVICES

Type of Motor	Percent of Full Load Current				
	Nontime Delay Fuse	Dual-Element (Time-Delay) Fuse	Instant. Trip Type Breaker	Time-Limit Breaker	
All A.C. single-phase and polyphase squirrel-cage and synchronous motors with full-voltage, resistance or reactor starting:					
No code letter	300	175	700	250	
Code letter F to V	300	175	700	250	
Code letter B to E	250	175	700	200	
Code letter A	150	150	700	150	
All A.C. squirrel-cage and synchronous motors with auto-transformer starting:					
Code letter F to V	250	175	700	200	
Code letter B to E	200	175	700	200	
Code letter A	150	150	700	150	
Wound Rotor	150	150	700	150	
Direct Current					
Not more than 50 HP	150	150	250	150	
More than 50 HP	150	150	175	150	

MAXIMUM OCPD

$$OCPD = FLC \times R_M$$

where

R_M = maximum rating of OCPD

and

FLC = full load current (from motor nameplate or NEC® Table 430.150)

Motor Type	Code Letter	FLC %				
		Motor Size	TDF	NTDF	ITB	ITCB
AC*	—	—	175	300	150	700
AC*	A	—	150	150	150	700
AC*	B-E	—	175	250	200	700
AC*	F-V	—	175	300	250	700
DC	—	⅛ TO 50 HP	150	150	150	250
DC	—	Over 50 HP	150	150	150	175

* full-voltage and resistor starting

STANDARD SIZES OF FUSES AND CB'S

NEC® 240.6 lists standard amperage ratings of fuses and fixed-trip circuit breakers as follows:

15	60	175	500	2500
20	70	200	600	3000
25	80	225	700	4000
30	90	250	800	5000
35	100	300	1000	6000
40	110	350	1200	
45	125	400	1600	
50	150	450	2000	

MOTOR POWER FORMULAS — COST SAVINGS

Power Consumed	Operating Cost	Annual Savings
$$P = \frac{HP \times 746}{E_{ff}}$$ where P = power consumed (W) HP = horsepower 746 = constant E_{ff} = efficiency (%)	$$C_{/hr} = \frac{P_{/hr} \times C_{/kWh}}{1000}$$ where $C_{/hr}$ = operating cost per hour $P_{/hr}$ = power consumed per hour $C_{/kWh}$ = cost per kilowatt hour 1000 = constant to remove kilo	$$S_{Ann} = C_{Ann\ Std} - C_{Ann\ Eff}$$ where S_{Ann} = annual cost savings $C_{Ann\ Std}$ = annual operating cost for standard motor $C_{Ann\ Eff}$ = annual operating cost for energy-efficient motor

HEATER SELECTIONS

Heater number	Full-load current (Amps)*				
	Size 0	Size 1	Size 2	Size 3	Size 4
10	.20	.20	—	—	—
11	.22	.22	—	—	—
12	.24	.24	—	—	—
13	.27	.27	—	—	—
14	.30	.30	—	—	—
15	.34	.34	—	—	—
16	.37	.37	—	—	—
17	.41	.41	—	—	—
18	.45	.45	—	—	—
19	.49	.49	—	—	—
20	.54	.54	—	—	—
21	.59	.59	—	—	—
22	.65	.65	—	—	—
23	.71	.71	—	—	—
24	.78	.78	—	—	—
25	.85	.85	—	—	—
26	.93	.93	—	—	—
27	1.02	1.02	—	—	—
28	1.12	1.12	—	—	—
29	1.22	1.22	—	—	—
30	1.34	1.34	—	—	—
31	1.48	1.48	—	—	—
32	1.62	1.62	—	—	—
33	1.78	1.78	—	—	—
34	1.96	1.96	—	—	—
35	2.15	2.15	—	—	—
36	2.37	2.37	—	—	—
37	2.60	2.60	—	—	—

*Full-load current (Amps) does not include FLC x 1.15 or 1.25.

HEATER SELECTIONS (cont.)

Heater number	Full-load current (Amps)*				
	Size 0	Size 1	Size 2	Size 3	Size 4
38	2.86	2.86	—	—	—
39	3.14	3.14	—	—	—
40	3.45	3.45	—	—	—
41	3.79	3.79	—	—	—
42	4.17	4.17	—	—	—
43	4.58	4.58	—	—	—
44	5.03	5.03	—	—	—
45	5.53	5.53	—	—	—
46	6.08	6.08	—	—	—
47	6.68	6.68	—	—	—
48	7.21	7.21	—	—	—
49	7.81	7.81	7.89	—	—
50	8.46	8.46	8.57	—	—
51	9.35	9.35	9.32	—	—
52	10.00	10.00	10.1	—	—
53	10.7	10.7	11.0	12.2	—
54	11.7	11.7	12.0	13.3	—
55	12.6	12.6	12.9	14.3	—
56	13.9	13.9	14.1	15.6	—
57	15.1	15.1	15.5	17.2	—
58	16.5	16.5	16.9	18.7	—
59	18.0	18.0	18.5	20.5	—
60	—	19.2	20.3	22.5	23.8
61	—	20.4	21.8	24.3	25.7
62	—	21.7	23.5	26.2	27.8
63	—	23.1	25.3	28.3	30.0
64	—	24.6	27.2	30.5	32.5
65	—	26.2	29.3	33.0	35.0

*Full-load current (Amps) does not include FLC x 1.15 or 1.25.

HEATER SELECTIONS (cont.)

Heater number	Full-load current (Amps)*				
	Size 0	Size 1	Size 2	Size 3	Size 4
66	—	27.8	31.5	36.0	38.0
67	—	—	33.5	39.0	41.0
68	—	—	36.0	42.0	44.5
69	—	—	38.5	38.5	48.5
70	—	—	41.0	49.5	52
71	—	—	43.0	43.0	57
72	—	—	46.0	58	61
73	—	—	—	63	67
74	—	—	—	68	72
75	—	—	—	73	77
76	—	—	—	78	84
77	—	—	—	83	91
78	—	—	—	88	97
79	—	—	—	—	103
80	—	—	—	—	111
81	—	—	—	—	119
82	—	—	—	—	127
83	—	—	—	—	133

*Full-load current (Amps) does not include FLC x 1.15 or 1.25.

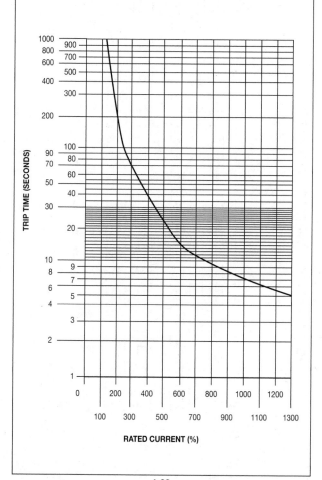

HEATER TRIP CHARACTERISTICS

TRIP TIME (SECONDS)

RATED CURRENT (%)

4-33

HEATER AMBIENT TEMPERATURE CORRECTION

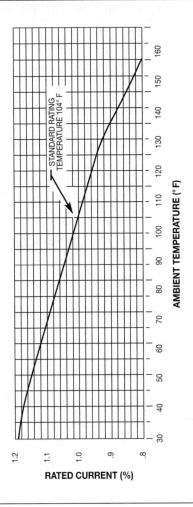

STANDARD RATING TEMPERATURE 104° F

RATED CURRENT (%)

AMBIENT TEMPERATURE (°F)

HEATING ELEMENT SPECIFICATIONS

Material State	Heated Material	Outer Cover Material	Maximum Watt Density	Material Operating Temperature
Liquid	Gasoline	Iron/steel	18-22	300
	Oil (low viscosity)	Steel/copper	20-25	Up to 180
	Oil (medium viscosity)	Steel/copper	12-18	Up to 180
	Oil (high viscosity)	Steel/copper	5-7	Up to 180
	Vegetable oil (cooking)	Copper	28-32	400
	Water (washroom)	Copper	70-90	140
	Water (process)	Copper	40-50	212
Air/Gas	Still air (ovens)	Steel/stainless steel	28-32 18-22 8-10 2-3	Up to 700 Up to 1000 Up to 1200 Up to 1500
	Air (moving at 10 fps)	Aluminum steel	30-34 23-26 14-16 2-3	Up to 700 Up to 1000 Up to 1200 Up to 1500
Solid	Asphalt	Iron/steel	9-12 8-10 6-8 5-6	Up to 200 Up to 300 Up to 400 Up to 500
	Molten tin (heated in pot)	Iron/steel	20-22	600
	Steel tubing (heated indirectly)	Iron/steel	45-48 50-52 54-56	500 750 1000

HEATER CONSTRUCTION

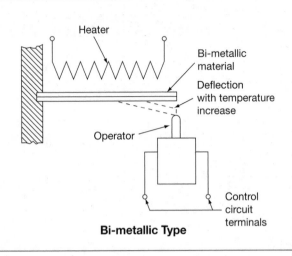

Bi-metallic Type

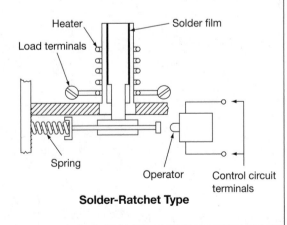

Solder-Ratchet Type

4-36

Test bearings by trying to move shaft vertically.

Check for gaps. If found, tap into place with mallet.

Gaps

MOTOR DISCONNECTS

For stationary motors rated at 2 HP or less and 300 volts or less, the disconnecting means may be a general-use switch with an ampere rating of at least the full-load current of the motor

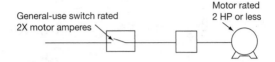

General-use switch rated 2X motor amperes

Motor rated 2 HP or less

For AC circuits only, an AC general-use snap switch may be used as a disconnecting means. Full load current may not exceed 80% of the ampere rating of the switch

---

For stationary motors rated at more than 100 HP, the disconnecting means may be a motor-circuit switch rated in amperes, a general-use switch or an isolating switch that must be marked "Do not open under load."

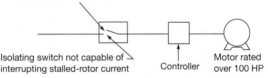

Isolating switch not capable of interrupting stalled-rotor current

Controller

Motor rated over 100 HP

A plug and receptacle may serve as the disconnecting means for portable motors if its HP rating is at least that of the motor or 1.4 times HP rating if Design E motor

Motor

MOTOR DISCONNECTS (cont.)

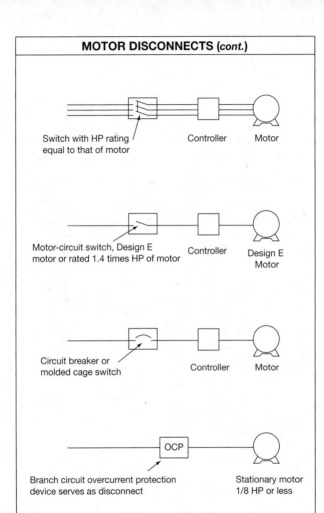

Switch with HP rating equal to that of motor — Controller — Motor

Motor-circuit switch, Design E motor or rated 1.4 times HP of motor — Controller — Design E Motor

Circuit breaker or molded cage switch — Controller — Motor

Branch circuit overcurrent protection device serves as disconnect — OCP — Stationary motor 1/8 HP or less

MOTOR FEEDER TAP SIZING

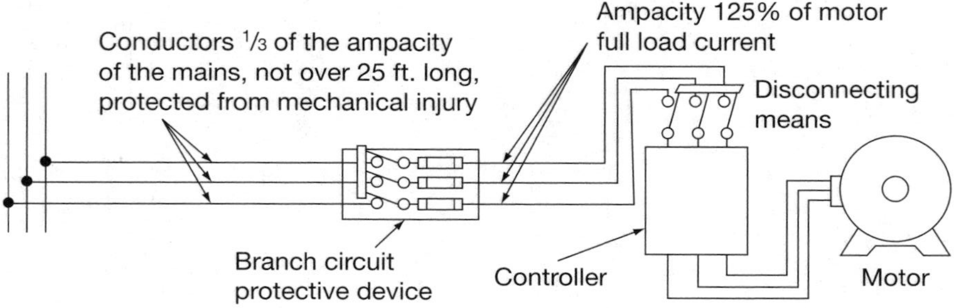

Ampacity 125% of motor full load current

Disconnecting means

Motor

Controller

Branch circuit protective device

Conductors 1/3 of the ampacity of the mains, not over 25 ft. long, protected from mechanical injury

A branch circuit fuse or circuit breaker may be rated higher than the ampacity of the tap conductors.

GENERAL CONTROLLER FAULT CURRENT LIMITS

HP rating	RMS (sym. amps)
1 or less	1,000
1½ to 50	5,000
51 to 200	10,000
201 to 400	18,000
401 to 600	30,000
601 to 900	42,000
901 to 1600	85,000

HORSEPOWER LIMITS OF INDIVIDUAL MOTORS

Voltage	Minimum HP	Maximum HP
115 DC	None	30
230 DC	None	200
550/600 DC	½	None
110 – 120 1φ AC	None	1½
220 – 240 1φ AC	None	10
440 – 550 1φ AC	5	10
110 – 120 2φ/3φ AC	None	15
220 – 240 2φ/3φ AC	None	200
440 – 550 2φ/3φ AC	None	500
2,200 2φ/3φ AC	40	None
4,000 2φ/3φ AC	75	None
6,600 2φ/3φ AC	400	None

COMMON SERVICE FACTORS

Equipment	Service factor
Blowers	
Centrifugal	1.00
Vane	1.25
Compressors	
Centrifugal	1.25
Vane	1.50
Conveyors	
Uniformly loaded or fed	1.50
Heavy-duty	2.00
Elevators	
Bucket	2.00
Freight	2.25
Extruders	
Plastic	2.00
Metal	2.50
Fans	
Light-duty	1.00
Centrifugal	1.50
Machine tools	
Bending roll	2.00
Punch press	2.25
Tapping machine	3.00
Mixers	
Concrete	2.00
Drum	2.25
Paper mills	
De-barking machines	3.00
Beater and pulper	2.00
Bleacher	1.00
Dryers	2.00
Log haul	2.00
Printing presses	1.50
Pumps	
Centrifugal—general	1.00
Centrifugal—sewage	2.00
Reciprocating	2.00
Rotary	1.50
Textile	
Batchers	1.50
Dryers	1.50
Looms	1.75
Spinners	1.50
Woodworking machines	1.00

SMALL MOTOR GUIDE

For AC type, 115 volt, 60 hz, single-phase

HP	Amps @ Full Load	RPM Speed
2	19.0 to 23.0	3450
1½	19.6	1725
	16.4 to 19.6	3450
1	13.6 to 16.0	1725
	13.0 to 15.0	3450
¾	9.5	1075
	11.6	1725
	11.8	3450
½	7.3	1075
	7.0 to 9.2	1725
	9.8	3450
⅓	5.1	1075
	5.0 to 7.2	1140
	5.3 to 6.8	1725
	5.6 to 6.5	3450
¼	6.9	850
	3.4 to 6.8	1075
	5.6 to 6.8	1140
	3.1 to 3.6	1625
	4.4 to 6.3	1725
⅙	2.4 to 5.0	1075
	4.0 to 4.9	1140
	4.0 to 4.8	1550
	3.3 to 4.7	1725
⅛	1.8 to 5.0	1075
	3.8	1140
	2.5	1725
1/10	4.0	1050
	3.5	1550
1/12	3.2	850
	4.1	1550
	2.8	1725
1/15	2.8	1550
1/20	2.5	1550

This chart is for small motors used in HVAC equip., fans, small pumps, etc. Specifications may vary per job requirement and motor used. When referencing 230 volt motors, divide the amperes by 2.

DC MOTOR PERFORMANCE CHARACTERISTICS

Performance Characteristics	Voltage 10% below Rated Voltage		Voltage 10% above Rated Voltage	
	Shunt	Compound	Shunt	Compound
Starting Torque	−15%	−15%	+15%	+15%
Speed	−5%	−6%	+5%	+6%
Current	+12%	+12%	−8%	−8%
Field temperature	Increases	Decreases	Increases	Increases
Armature temperature	Increases	Increases	Decreases	Decreases
Commutator temperature	Increases	Increases	Decreases	Decreases

MAXIMUM ACCELERATION TIME

Motor Frame Number	Maximum Acceleration Time
48 and 56	8 seconds
143-286	10 seconds
324-326	12 seconds
364-505	15 seconds

DC MOTOR EFFICIENCIES, CONTINUOUS RATED, 40°C RISE

HP	Efficiency Percentage			HP	Efficiency Percentage		
	Load	Load	Full Load		Load	Load	Full Load
½	62.0	65.0	68.0	25	83.0	86.0	87.0
¾	65.0	70.0	72.0	30	84.0	87.0	88.0
1	70.0	73.0	75.0	40	85.0	87.5	88.5
1½	72.0	76.0	78.0	50	85.5	88.0	89.5
2	72.0	77.5	81.0	60	85.5	88.0	90.0
3	72.0	77.0	79.0	75	86.0	89.0	90.5
5	77.0	81.0	81.5	100	86.0	89.0	90.5
7½	79.0	82.0	83.5	125	86.0	89.0	90.5
10	81.0	83.0	85.0	150	87.0	90.0	91.0
15	81.5	84.5	86.0	200	89.0	91.5	92.0
20	82.0	85.0	86.5				

POWER FACTOR IMPROVEMENT
Capacitor Multipliers for Kilowatt Load

To give capacitor kvar required to improve power-factor from original to desired value. For example, assume the total plant load is 100kw at 60 percent power factor. Capacitor kvar rating necessary to improve power factor to 80 percent is found by multiplying kw (100) by multiplier in table (0.583), which gives kvar (58.3). Nearest standard rating (60 kvar) should be recommended.

Original Power Factor Percentage	Desired Power Factor - Percentage				
	100%	95%	90%	85%	80%
60	1.333	1.004	0.849	0.713	0.583
62	1.266	0.937	0.782	0.646	0.516
64	1.201	0.872	0.717	0.581	0.451
66	1.138	0.809	0.654	0.518	0.388
68	1.078	0.749	0.594	0.458	0.338
70	1.020	0.691	0.536	0.400	0.270
72	0.964	0.635	0.480	0.344	0.214
74	0.909	0.580	0.425	0.289	0.159
76	0.855	0.526	0.371	0.235	0.105
77	0.829	0.500	0.345	0.209	0.079
78	0.802	0.473	0.318	0.182	0.052
79	0.776	0.447	0.292	0.156	0.026
80	0.750	0.421	0.266	0.130	
81	0.724	0.395	0.240	0.104	
82	0.698	0.369	0.214	0.078	
83	0.672	0.343	0.188	0.052	
84	0.646	0.317	0.162	0.206	
85	0.620	0.291	0.136		
86	0.593	0.264	0.109		
87	0.567	0.238	0.083		
88	0.540	0.211	0.056		
89	0.512	0.183	0.028		
90	0.484	0.155			
91	0.456	0.127			
92	0.426	0.097			
93	0.395	0.066			
94	0.363	0.034			
95	0.329				
96	0.292				
97	0.251				
99	0.143				

TYPICAL MOTOR POWER FACTORS

HP	Speed (RPM)	Power Factor at		
		½ load	¾ load	full load
0 – 5	1800	.72	.82	.84
5.01 – 20	1800	.74	.84	.86
20.1 – 100	1800	.79	.86	.89
100.1 – 300	1800	.81	.88	.91

EFFICIENCY FORMULAS

Input and Output Power Known	Horsepower and Power Loss Known
$$E_{ff} = \frac{P_{out}}{P_{in}}$$ where E_{ff} = efficiency (%) P_{out} = output power (W) P_{in} = input power (W)	$$E_{ff} = \frac{746 \times HP}{746 \times HP + W_l}$$ where E_{ff} = efficiency (%) 746 = constant HP = horsepower W_l = watts lost

VOLTAGE UNBALANCE FORMULA

$$V_u = \frac{V_d}{V_a} \times 100$$

where
V_u = voltage unbalance (%)
V_d = voltage deviation (V)
V_a = voltage average (V)
100 = constant

MOTOR TORQUE (INCH POUNDS-FORCE)

$$\frac{63025 \times HP}{R.P.M.} = \text{Torque in inch pounds - force}$$

$$\left(\begin{array}{c} \text{Divide the above by 12} \\ \text{to obtain torque in foot pounds} \end{array} \right)$$

Motor	RPM Speeds of Motors			
HP	68	100	155	190
600	556103	378150	243968	199026
550	509761	346638	223637	182441
500	463419	315125	203306	165855
450	417077	283613	182976	149270
400	370735	252100	162645	132684
350	324393	220588	142315	116099
300	278051	189075	121984	99513
275	254881	173319	111819	91220
250	231710	157563	101653	82928
225	208539	141806	91488	74635
200	185368	126050	81323	66342
175	162197	110294	71157	58049
150	139026	94538	60992	49757
125	115855	78781	50827	41464
100	92684	63025	40661	33171
90	83415	56723	36595	29854
80	74147	50420	32529	26537
70	64879	44118	28463	23220
60	55610	37815	24397	19903
50	46342	31513	20331	16586
40	37074	25210	16265	13268
30	27805	18908	12198	9951
25	23171	15756	10165	8293
20	18537	12605	8132	6634
15	13903	9454	6099	4976
10	9268	6303	4066	3317
7½	6951	4727	3050	2488
5	4634	3151	2033	1659
3	2781	1891	1220	995
2	1854	1261	813	663
1½	1390	945	610	498
1	927	630	407	332

MOTOR TORQUE (INCH POUNDS-FORCE) (cont.)

$$\frac{63025 \times HP}{R.P.M.} = \text{Torque in inch pounds - force}$$

$$\left(\begin{array}{c} \text{Divide the above by 12} \\ \text{to obtain torque in foot pounds} \end{array} \right)$$

Motor	RPM Speeds of Motors			
HP	500	750	850	1000
600	75630	50420	44488	37815
550	69328	46218	40781	34664
500	63025	42017	37074	31513
450	56723	37815	33366	28361
400	50420	33613	29659	25210
350	44118	29412	25951	22059
300	37815	25210	22244	18908
275	34664	23109	20390	17332
250	31513	21008	18537	15756
225	28361	18908	16683	14181
200	25210	16807	14829	12605
175	22059	14706	12976	11029
150	18908	12605	11122	9454
125	15756	10504	9268	7878
100	12605	8403	7415	6303
90	11345	7563	6673	5672
80	10084	6723	5932	5042
70	8824	5882	5190	4412
60	7563	5042	4449	3782
50	6303	4202	3707	3151
40	5042	3361	2966	2521
30	3782	2521	2224	1891
25	3151	2101	1854	1576
20	2521	1691	1483	1261
15	1891	1261	1112	945
10	1261	840	741	630
7½	945	630	556	473
5	630	420	371	315
3	378	252	222	189
2	252	168	148	126
1½	189	126	111	95
1	126	84	74	63

MOTOR TORQUE (INCH POUNDS-FORCE) (cont.)

$$\frac{63025 \times HP}{R.P.M.} = \text{Torque in inch pounds - force}$$

$$\left(\begin{array}{c} \text{Divide the above by 12} \\ \text{to obtain torque in foot pounds} \end{array} \right)$$

Motor	RPM Speeds of Motors			
HP	1050	1550	1725	3450
600	36014	24397	21922	10961
550	33013	22364	20095	10047
500	30012	20331	18268	9134
450	27011	18298	16441	8221
400	24010	16265	14614	7307
350	21008	14231	12788	6394
300	18007	12198	10961	5480
275	16507	11182	10047	5024
250	15006	10165	9134	4567
225	13505	9149	8221	4110
200	12005	8132	7307	3654
175	10504	7116	6394	3197
150	9004	6099	5480	2740
125	7503	5083	4567	2284
100	6002	4066	3654	1827
90	5402	3660	3288	1644
80	4802	3253	2923	1461
70	4202	2846	2558	1279
60	3601	2440	2192	1096
50	3001	2033	1827	913
40	2401	1626	1461	731
30	1801	1220	1096	548
25	1501	1017	913	457
20	1200	813	731	365
15	900	610	548	274
10	600	407	365	183
7½	450	305	274	137
5	300	203	183	91
3	180	122	110	55
2	120	81	73	37
1½	90	61	55	27
1	60	41	37	18

HORSEPOWER TO TORQUE CONVERSION

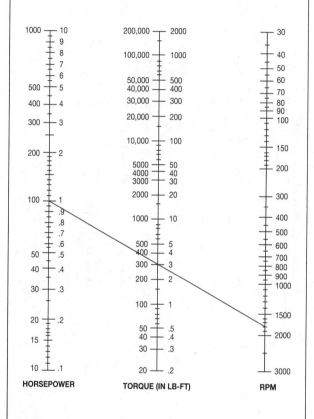

HORSEPOWER   **TORQUE (IN LB-FT)**   **RPM**

LOCKED ROTOR CURRENT

Apparent, 1φ	Apparent, 3φ	True, 1φ	True, 3φ
$LRC = \dfrac{1000 \times HP \times kVA/HP}{V}$	$LRC = \dfrac{1000 \times HP \times kVA/HP}{V \times \sqrt{3}}$	$LRC = \dfrac{1000 \times HP \times kVA/HP}{V \times PF \times E_{ff}}$	$LRC = \dfrac{1000 \times HP \times kVA/HP}{V \times \sqrt{3} \times PF \times E_{ff}}$
where	where	where	where
LRC = locked rotor current (in amps)	LRC = locked rotor current (in amps)	LRC = locked rotor current (in amps)	LRC = locked rotor current (in amps)
1000 = multiplier for kilo	1000 = multiplier for kilo	1000 = multiplier for kilo	1000 = multiplier for kilo
HP = horsepower	HP = horsepower	HP = horsepower	HP = horsepower
kVA/HP = kilovolt amps per horsepower	kVA/HP = kilovolt amps per horsepower	kVA/HP = kilovolt amps per horsepower	kVA/HP = kilovolt amps per horsepower
V = volts	V = volts	V = volts	V = volts
	$\sqrt{3} = 1.732$	PF = power factor	$\sqrt{3} = 1.732$
		E_{ff} = motor efficiency	

NEMA RATINGS OF 60HZ AC CONTACTORS

Size	8 Hr Open Rating (in Amps)	Power in Volts				
		3φ			1φ	
		200	230	230/460	115	230
00	9	1½	1½	2	⅓	1
0	18	3	3	5	1	2
1	27	7½	7½	10	2	3
2	45	10	15	25	3	7½
3	90	25	30	50	–	–
4	135	40	50	100	–	–
5	270	75	100	200	–	–
6	540	150	200	400	–	–
7	810	–	300	600	–	–
8	1215	–	450	900	–	–
9	2250	–	800	1600	–	–

FULL-LOAD CURRENTS: DC MOTORS

HP (Motor rating)	Current (Amps)	
	120 V	240 V
¼	3.1	1.6
⅓	4.1	2.0
½	5.4	2.7
¾	7.6	3.8
1	9.5	4.7
1½	13.2	6.6
2	17	8.5
3	25	12.2
5	40	20
7½	48	29
10	76	38

FULL-LOAD CURRENTS: 1ϕ, AC MOTORS

HP (Motor rating)	Current (Amps)	
	115 V	230 V
⅙	4.4	2.2
¼	5.8	2.9
⅓	7.2	3.6
½	9.8	4.9
¾	18.8	6.9
1	16	8
1½	20	10
2	24	12
3	34	17
5	56	28
7½	80	40
10	100	50

FULL-LOAD CURRENTS:
3φ, AC INDUCTION MOTORS

HP (Motor rating)	Current (Amps)			
	208 V	230 V	460 V	575 V
¼	1.11	.96	.48	.38
⅓	1.34	1.18	.59	.47
½	2.2	2.0	1.0	.8
¾	3.1	2.8	1.4	1.1
1	4.0	3.6	1.8	1.4
1½	5.7	5.2	2.6	2.1
2	7.5	6.8	3.4	2.7
3	10.6	9.6	4.8	3.9
5	16.7	15.2	7.6	6.1
7½	24.0	22.0	11.0	9.0
10	31.0	28.0	14.0	11.0
15	46.0	42.0	21.0	17.0
20	59	54	27	22
25	75	68	34	27
30	88	80	40	32
40	114	104	52	41
50	143	130	65	52
60	169	154	77	62
75	211	192	96	77
100	273	248	124	99
125	343	312	156	125
150	396	360	180	144
200	—	480	240	192
250	—	602	301	242
300	—	—	362	288
350	—	—	413	337
400	—	—	477	382
500	—	—	590	472

GENERAL EFFECT OF VOLTAGE VARIATION ON DIRECT CURRENT MOTOR CHARACTERISTICS

□ =INCREASE ● =DECREASE

Voltage Variation	Starting and Max. Run Torque	Full-load Speed	EFFICIENCY			Full-load Current	Temperature Rise, Full Load	Maximum Overload Capacity	Magnetic Noise
			Full Load	3/4 Load	1/2 Load				
SHUNT-WOUND									
120% Voltage	□ 30%	□ 110%	Slight □	No Change	Slight ●	● 17%	Main field □ Commutating field and armature ●	□ 30%	Slight □
110% Voltage	□ 15%	□ 105%	Slight □	No Change	Slight ●	● 8.5%	Main field □ Commutating field and armature ●	□ 15%	Slight □
90% Voltage	● 16%	● 95%	Slight ●	No Change	Slight □	● 11.5%	Main field ● Commutating field and armature ●	● 16%	Slight ●
COMPOUND-WOUND									
120% Voltage	□ 30%	□ 112%	Slight □	No Change	Slight ●	● 17%	Main field □ Commutating field and armature ●	□ 30%	Slight □
110% Voltage	□ 15%	□ 106%	Slight □	No Change	Slight ●	● 8.5%	Main field □ Commutating field and armature ●	□ 15%	Slight □
90% Voltage	● 16%	● 94%	Slight ●	No Change	Slight □	□ 11.5%	Main field ● Commutating field and armature □	● 16%	Slight ●

NOTES:—Starting current is controlled by starting resistor.
This table shows general effects, which will vary somewhat for specific ratings.

GENERAL EFFECT OF VOLTAGE AND FREQUENCY VARIATION ON INDUCTION MOTOR CHARACTERISTICS

NOTE: This table shows general effects, which will vary somewhat for specific ratings.

□ = INCREASE   ● = DECREASE

		Starting and Max. Running Torque	Synchronous Speed	% Slip	Full-load Speed	EFFICIENCY Full Load	EFFICIENCY 3/4 Load	EFFICIENCY 1/2 Load	POWER-FACTOR Full Load	POWER-FACTOR 3/4 Load	POWER-FACTOR 1/2 Load	Full-load Current	Starting Current	Temperature Rise, Full Load	Maximum Overload Capacity	Magnetic Noise, No Load in Particular
Voltage Variation	120% Voltage	□ 44%	No Change	● 30%	□ 1.5%	Small □	□ 1/2 to 2 points	● 7 to 20 points	● 5 to 15 points	● 10 to 30 points	● 15 to 40 points	● 11%	□ 25%	● 5 to 6C	□ 44%	Noticeable □
	110% Voltage	□ 21%	No Change	● 17%	□ 1%	□ 1/2 to 1 point	Practically no change	● 1 to 2 points	● 3 points	● 4 points	● 5 to 6 points	● 7%	□ 10 to 12%	● 3 to 4C	□ 21%	Slight □
	Function of Voltage	$(\text{Voltage})^2$	Constant	$\dfrac{1}{(\text{Voltage})^2}$	(Syn speed – slip)	—	—	—	—	—	—	—	Voltage	—	$(\text{Voltage})^2$	—
	90% Voltage	● 19%	No Change	□ 23%	□ 1½%	● 2 points	Practically no change	□ 1 to 2 points	□ 1 point	□ 2 to 3 points	□ 4 to 5 points	□ 11%	● 10 to 12%	□ 6 to 7C	● 19%	Slight ●
Frequency Variation	105% Frequency	● 10%	□ 5%	Practically no change	□ 5%	Slight □	Slight □	Slight □	Slight ●	Slight ●	Slight ●	Slight ●	● 5 to 6%	Slight ●	—	—
	Function of Frequency	$\dfrac{1}{(\text{Frequency})^2}$	Frequency	—	(Syn speed – slip)	—	—	—	—	—	—	—	$\dfrac{1}{(\text{Frequency})}$	—	—	—
	95% Frequency	□ 11%	● 5%	Practically no change	● 5%	Slight ●	Slight ●	Slight ●	Slight □	Slight □	Slight □	Slight □	□ 5 to 6%	Slight □	Slight □	Slight □

VOLTAGE VARIATION CHARACTERISTICS

Performance Characteristics	10% above Rated Voltage	10% below Rated Voltage
Starting current	+10% to +12%	−10% to −12%
Full-load current	−7%	+11%
Motor torque	+20% to +25%	−20% to −25%
Motor efficiency	Little change	Little change
Speed	+1%	-1.5%
Temperature rise	−3°C to −4°C	+6°C to +7°C

FREQUENCY VARIATION CHARACTERISTICS

Performance Characteristics	5% above Rated Frequency	5% below Rated Frequency
Starting current	−5% to −6%	+5% to +6%
Full-load current	−1%	+1%
Motor torque	−10%	+11%
Motor efficiency	Slight increase	Slight decrease
Speed	−5%	−5%
Temperature rise	Slight decrease	Slight increase

TYPICAL MOTOR EFFICIENCIES

HP	Standard Motor (%)	Energy-Efficient Motor (%)
1	76.5	84.0
1.5	78.5	85.5
2	79.9	86.5
3	80.8	88.5
5	83.1	88.6
7.5	83.8	90.2
10	85.0	90.3
15	86.5	91.7
20	87.5	92.4
25	88.0	93.0
30	88.1	93.1
40	89.3	93.6
50	90.4	93.7
75	90.8	95.0
100	91.6	95.4
125	91.8	95.8
150	92.3	96.0
200	93.3	96.1
250	93.6	96.2
300	93.8	96.5

TYPES OF ENCLOSURES

OPEN-TYPE
has full openings in
frame and endbells for
maximum ventilation, is
lowest cost enclosure.

SEMI-PROTECTED
has screens in top
openings to keep
out falling objects.
PROTECTED has
screens in bottom too.

DRIP-PROOF
has upper parts covered
to keep out drippings
falling at angle not
over 15° from vertical

SPLASH-PROOF
is baffled at bottom to
keep out particles
coming at angle not
over 100° from vertical

TOTALLY-ENCLOSED
can be non-ventilated,
separately ventilated,
or explosion proof for
hazardous atmospheres

FAN-COOLED
totally-enclosed motor
has double covers. Fan,
behind vented outer
shroud, is run by motor

ENCLOSURE INSULATION, TEMPERATURE

CLASS A (cotton, silk, paper or other
organics impregnated with insulating
varnish) is considered standard for
most applications, allows 105° C
total temperature

  40°C ambient
  40°C rise by thermometer
  15°C "hot-spot" allowance
  10°C service factor
  105°C total temperature

CLASS B (mica, asbestos, fiber-glass,
other inorganics) allows 130°C total
temperature

  40°C ambient
  70°C rise by thermometer
  20°C "hot-spot" allowance
  130°C total temperature

CLASS H (including silicone family) is
for special high-temperature
applications

SPECIAL CLASS A is highly resistant,
but not "proof," against severe
moisture, dampness; conduc-
tive, corrosive or abrasive dusts
and vapors

TROPICAL is for excessive moisture,
high ambients, corrosion, fungus,
vermin, insects

ENCLOSURE RATINGS

Type	Service Conditions	Sealing Method	Cost
1	No unusual		Base
4	Windblown dust and rain, splashing water, hose-directed water, and ice on enclosure		12 x Base
4X	Corrosion, windblown dust and rain, splashing water, hose-directed water, and ice on enclosure.		12 x Base
7	Withstand and contain an internal explosion of specified gases, contain an explosion sufficiently so an explosive gas-air mixture in the atmosphere is not ignited.		48 x Base
9	Dust		48 x Base
12	Dust, falling dirt, and dripping noncorrosive liquids		5 x Base

HAZARDOUS LOCATIONS

Class	Group	Material
I	A	Acetylene
	B	Hydrogen, butadiene, ethylene oxide, propylene oxide
	C	Carbon monoxide, ether, ethylene, hydrogen sulfide, morpholine, cyclopropane
	D	Gasoline, benzene, butane, propane, alcohol, acetone, ammonia, vinyl chloride
II	E	Metal dusts
	F	Carbon black, coke dust, coal
	G	Grain dust, flour, starch, sugar, plastics
III	No groups	Wood chips, cotton, flax, and nylon

ENCLOSURE TYPES

Type	Use	Service Conditions	UL Tests	Comments
1	Indoor	None	Rod entry, rust resistance	
3	Outdoor	Windblown dust, rain, sleet, and ice on enclosure	Rain, external icing, dust, and rust resistance	Do not provide protection against internal condensation, or internal icing
3R	Outdoor	Falling rain and ice on enclosure	Rod entry, rain, external icing, and rust resistance	Do not provide protection against dust, internal condensation, or internal icing
4	Indoor/outdoor	Windblown dust and rain, splashing water, hose-directed water and ice on enclosure	Hosedown, external icing and rust resistance	Do not provide protection against internal condensation, or internal icing
4X	Indoor/outdoor	Corrosion, windblown dust and rain, splashing water, hose-directed water and ice on enclosure	Hosedown, external icing, and corrosion resistance	Do not provide protection against internal condensation, or internal icing
6	Indoor/outdoor	Occasional temporary submersion at a limited depth		
6P	Indoor/outdoor	Prolonged submersion at a limited depth		

ENCLOSURE TYPES (cont.)

Type	Use	Service Conditions	UL Tests	Comments
7	Indoor locations classified as Class I, or Groups A, B, C, or D, as defined in the NEC®	Withstand and contain an internal explosion of specified gases, contain an explosion sufficiently so an explosive gas-air mixture in the atmospheres not ignited	Explosion, hydrostatic, and temperature	Enclosed heat-generating devices shall not cause external surfaces to reach temperatures capable of igniting explosive gas-air mixtures in the atmosphere
9	Indoor locations classified as Class II, Groups E or G, as defined in the NEC®	Dust	Dust penetration, temperature, and gasket aging	Enclosed heat-generating devices shall not cause external surfaces to reach temperatures capable of igniting explosive gas-air mixtures in the atmosphere
12	Indoor	Dust, falling dirt, and dripping noncorrosive liquids	Drip, dust, and rust resistance	Do not provide protection against internal condensation
13	Indoor	Dust, spraying water, oil and noncorrosive coolant	Oil explosion and rust resistance	Do not provide protection against internal condensation

FRONTAL VIEW OF TYPICAL MOTOR

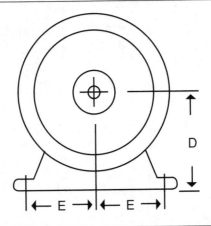

REFERENCE PAGE 4-65 FOR DIMENSIONS

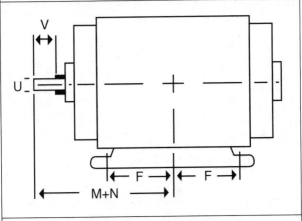

SIDE VIEW OF TYPICAL MOTOR

NEMA MOTOR FRAME DIMENSIONS

Frame No.	Dimensions in Inches							
	U	V	Keyway		D	E	F	M+N
42	3/8				2 5/8	1 3/4	27/32	4 1/32
48	1/2				2	2 1/8	1 3/8	5 3/8
56	5/8		3/16	3/32	3 1/2	2 7/16	1 1/2	6 1/8
66	3/4		3/16	3/32	4 1/8	2 15/16	2 1/2	7 7/8
143T	7/8	2	3/16	3/32	3 1/2	2 3/4	2	6 1/2
145T	7/8	2	3/16	3/32	3 1/2	2 3/4	2 1/2	7
182	7/8	2	3/16	3/32	4 1/2	3 3/4	2 1/4	7 1/4
182T	1 1/8	2 1/2	1/4	1/8	4 1/2	3 3/4	2 1/4	7 3/4
184	7/8	2	3/16	3/32	4 1/2	3 3/4	2 3/4	7 3/4
184T	1 1/8	2 1/2	1/4	1/8	4 1/2	3 3/4	2 3/4	8 1/4
213	1 1/8	2 3/4	1/4	1/8	5 1/4	4 1/4	2 3/4	9 1/4
213T	1 3/8	3 1/8	5/16	5/32	5 1/4	4 1/4	2 3/4	9 5/8
215	1 1/8	2 3/4	1/4	1/8	5 1/4	4 1/4	3 1/2	10
215T	1 3/8	3 1/8	5/16	5/32	5 1/4	4 1/4	3 1/2	10 3/8
254T	1 5/8	3 3/4	3/8	3/16	6 1/4	5	4 1/8	12 3/8
254U	1 3/8	3 1/2	5/16	5/32	6 1/4	5	4 1/8	12 1/8
256T	1 5/8	3 3/4	3/8	3/16	6 1/4	5	5	13 1/4
256U	1 3/8	3 1/2	5/16	5/32	6 1/4	5	5	13
284T	1 7/8	4 3/8	1/2	1/4	7	5 1/2	4 3/8	14 1/8
284TS	1 5/8	3	3/8	3/16	7	5 1/2	4 3/4	12 3/4
284U	1 5/8	4 5/8	3/8	3/16	7	5 1/2	4 3/4	14 3/8
286T	1 7/8	4 3/8	1/2	1/4	7	5 1/2	5 1/2	14 7/8
286U	1 5/8	4 5/8	3/8	3/16	7	5 1/2	5 1/2	15 1/8
324T	2 1/8	5	1/2	1/4	8	6 1/4	5 1/4	15 3/4
324U	1 7/8	5 3/8	1/2	1/4	8	6 1/4	5 1/4	16 1/8
326T	2 1/8	5	1/2	1/4	8	6 1/4	6	16 1/2
326TS	1 7/8	3 1/2	1/2	1/4	8	6 1/4	6	15
326U	1 7/8	5 3/8	1/2	1/4	8	6 1/4	6	16 7/8
364T	2 3/8	5 5/8	5/8	5/16	9	7	5 5/8	17 3/8
364U	2 1/8	6 1/8	1/2	1/4	9	7	5 5/8	17 7/8
365T	2 3/8	5 5/8	5/8	5/16	9	7	6 1/8	17 7/8
365U	2 1/8	6 1/8	1/2	1/4	9	7	6 1/8	18 3/8
404T	2 7/8	7	3/4	3/8	10	8	6 1/8	20
404U	2 3/8	6 7/8	5/8	5/16	10	8	6 1/8	19 7/8
405T	2 7/8	7	3/4	3/8	10	8	6 7/8	20 3/4
405U	2 3/8	6 7/8	5/8	5/16	10	8	6 7/8	20 5/8
444T	3 3/8	8 1/4	7/8	7/16	11	9	7 1/4	23 1/4
444U	2 7/8	8 3/8	3/4	3/8	11	9	7 1/4	23 3/8
445T	3 3/8	8 1/4	7/8	7/16	11	9	8 1/4	24 1/4
445U	2 7/8	8 3/8	3/4	3/8	11	9	8 1/4	24 3/8

Standards established by National Electrical Manufactures Association.

SHAFT COUPLING SELECTIONS

Coupling number	Rated torque (lb-in)	Maximum shock torque (lb-in)
10-101-A	16	45
10-102-A	36	100
10-103-A	80	220
10-104-A	132	360
10-105-A	176	480
10-106-A	240	660
10-107-A	325	900
10-108-A	525	1450
10-109-A	875	2450
10-110-A	1250	3500
10-111-A	1800	5040
10-112-A	2200	6160

MOTOR FRAME LETTERS

Letter	Designation
G	Gasoline pump motor
K	Sump pump motor
M and N	Oil burner motor
S	Standard short shaft for direct connection
T	Standard dimensions established
U	Previously used as frame designation for which standard dimensions are established
Y	Special mounting dimensions required from manufacturer
Z	Standard mounting dimensions except shaft extension

REFERENCE PAGE 4-68 AND 4-69 FOR DIMENSIONS

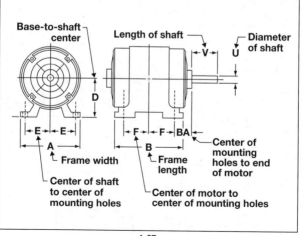

Base-to-shaft center

Length of shaft — V

Diameter of shaft — U

D

E + E

A — Frame width

Center of shaft to center of mounting holes

F + F + BA

B — Frame length

Center of motor to center of mounting holes

Center of mounting holes to end of motor

MOTOR FRAME DIMENSIONS

Frame No.	U	V	Keyway			A	B	D	E	F	BA
48	1/2	1 1/2*	flat	3/64	—	5 5/8*	3 1/2*	3	2 1/8	1 3/8	2 1/2
56	5/8	1 7/8*	3/16	3/16	1 3/8	6 1/2*	4 1/4*	3 1/2	2 7/16	1 1/2	2 3/4
143T	7/8	2	3/16	3/16	1 3/8	7	6	3 1/2	2 3/4	2	2 1/4
145T	7/8	2	3/16	3/16	1 3/8	7	6	3 1/2	2 3/4	2 1/2	2 1/4
182	7/8	2	3/16	3/16	1 3/8	9	6 1/2	4 1/2	3 3/4	2 1/4	2 3/4
182T	1 1/8	2 1/2	1/4	1/4	1 3/4	9	6 1/2	4 1/2	3 3/4	2 1/4	2 3/4
184	7/8	2	3/16	3/16	1 3/8	9	7 1/2	4 1/2	3 3/4	2 3/4	2 3/4
184T	1 1/8	2 1/2	1/4	1/4	1 3/4	9	7 1/2	4 1/2	3 3/4	2 3/4	2 3/4
203	3/4	2	3/16	3/16	1 3/8	10	7 1/2	5	4	2 3/4	3 1/8
204	3/4	2	3/16	3/16	1 3/8	10	8 1/2	5	4	3 1/4	3 1/8
213	1 1/8	2 3/4	1/4	1/4	2	10 1/2	7 1/2	5 1/4	4 1/4	2 3/4	3 1/2
213T	1 3/8	3 1/4	5/16	5/16	2 3/8	10 1/2	7 1/2	5 1/4	4 1/4	2 3/4	3 1/2
215	1 1/8	2 3/4	1/4	1/4	2	10 1/2	9	5 1/4	4 1/4	3 1/2	3 1/2
215T	1 3/8	3 1/8	5/16	5/16	2 3/8	10 1/2	9	5 1/4	4 1/4	3 1/2	3 1/2
224	1	2 3/4	1/4	1/4	2	11	8 3/4	5 1/2	4 1/2	3 3/8	3 1/2
225	1	2 3/4	1/4	1/4	2	11	9 1/2	5 1/2	4 1/2	3 3/4	3 1/2
254	1 1/8	3 1/8	1/4	1/4	2 3/8	12 1/2	10 3/4	6 1/4	5	4 1/8	4 1/4
254U	1 3/8	3 1/2	5/16	5/16	2 3/4	12 1/2	10 3/4	6 1/4	5	4 1/8	4 1/4
254T	1 5/8	3 3/4	3/8	3/8	2 7/8	12 1/2	10 3/4	6 1/4	5	4 1/8	4 1/4
256U	1 3/8	3 1/2	5/16	5/16	2 3/4	12 1/2	12 1/2	6 1/4	5	5	4 1/4
256T	1 5/8	3 3/4	3/8	3/8	2 7/8	12 1/2	12 1/2	6 1/4	5	5	4 1/4
284	1 1/4	3 1/2	1/4	1/4	2 3/4	14	12 1/2	7	5 1/2	4 3/4	4 3/4
284U	1 5/8	4 5/8	3/8	3/8	3 3/4	14	12 1/2	7	5 1/2	4 3/4	4 3/4
284T	1 7/8	4 3/8	1/2	1/2	3 1/4	14	12 1/2	7	5 1/2	4 3/4	4 3/4
284TS	1 5/8	3	3/8	3/8	1 7/8	14	12 1/2	7	5 1/2	4 3/4	4 3/4
286U	1 5/8	4 5/8	3/8	3/8	3 3/4	14	14	7	5 1/2	5 1/2	4 3/4
286T	1 7/8	4 3/8	1/2	1/2	3 1/4	14	14	7	5 1/2	5 1/2	4 3/4
286TS	1 5/8	3	3/8	3/8	1 7/8	14	14	7	5 1/2	5 1/2	4 3/4
324	1 5/8	4 5/8	3/8	3/8	3 3/4	16	14	8	6 1/4	5 1/4	5 1/4
324U	1 7/8	5 3/8	1/2	1/2	4 1/4	16	14	8	6 1/4	5 1/4	5 1/4
324S	1 5/8	3	3/8	3/8	1 7/8	16	14	8	6 1/4	5 1/4	5 1/4
324T	2 1/8	5	1/2	1/2	3 7/8	16	14	8	6 1/4	5 1/4	5 1/4
324TS	1 7/8	3 1/2	1/2	1/2	2	16	14	8	6 1/4	5 1/4	5 1/4
326	1 5/8	4 5/8	3/8	3/8	3 3/4	16	15 1/2	8	6 1/4	6	5 1/4
326U	1 7/8	5 3/8	1/2	1/2	4 1/4	16	15 1/2	8	6 1/4	6	5 1/4
326S	1 5/8	3	3/8	3/8	1 7/8	16	15 1/2	8	6 1/4	6	5 1/4
326T	2 1/8	5	1/2	1/2	3 7/8	16	15 1/2	8	6 1/4	6	5 1/4
326TS	1 7/8	3 1/2	1/2	1/2	2	16	15 1/2	8	6 1/4	6	5 1/4

*Not NEMA standard dimensions

MOTOR FRAME DIMENSIONS (cont.)

Frame No.	U	V	Keyway			A	B	D	E	F	BA
364	1 7/8	5 3/8	1/2	1/2	4 1/4	18	15 1/4	9	7	5 5/8	5 7/8
364S	1 5/8	3	3/8	3/8	1 7/8	18	15 1/4	9	7	5 5/8	5 7/8
364U	2 1/8	6 1/8	1/2	1/2	5	18	15 1/4	9	7	5 5/8	5 7/8
364US	1 7/8	3 1/2	1/2	1/2	2	18	15 1/4	9	7	5 5/8	5 7/8
364T	2 3/8	5 5/8	5/8	5/8	4 1/4	18	15 1/4	9	7	5 5/8	5 7/8
364TS	1 7/8	3 1/2	1/2	1/2	2	18	15 1/4	9	7	5 5/8	5 7/8
365	1 7/8	5 3/8	1/2	1/2	4 1/4	18	16 1/4	9	7	6 1/8	5 7/8
365S	1 5/8	3	3/8	3/8	1 7/8	18	16 1/4	9	7	6 1/8	5 7/8
365U	2 1/8	6 1/8	1/2	1/2	5	18	16 1/4	9	7	6 1/8	5 7/8
365US	1 7/8	3 1/2	1/2	1/2	2	18	16 1/4	9	7	6 1/8	5 7/8
365T	2 3/8	5 5/8	5/8	5/8	4 1/4	18	16 1/4	9	7	6 1/8	5 7/8
365TS	1 7/8	3 1/2	1/2	1/2	2	18	16 1/4	9	7	6 1/8	5 7/8
404	2 1/8	6 1/8	1/2	1/2	5	20	16 1/4	10	8	6 1/8	6 5/8
404S	1 7/8	3 1/2	1/2	1/2	2	20	16 1/4	10	8	6 1/8	6 5/8
404U	2 3/8	6 7/8	5/8	5/8	5 1/2	20	16 1/4	10	8	6 1/8	6 5/8
404US	2 1/8	4	1/2	1/2	2 3/4	20	16 1/4	10	8	6 1/8	6 5/8
404T	2 7/8	7	3/4	3/4	5 5/8	20	16 1/4	10	8	6 1/8	6 5/8
404TS	2 1/8	4	1/2	1/2	2 3/4	20	16 1/4	10	8	6 1/8	6 5/8
405	2 1/8	6 1/8	1/2	1/2	5	20	17 3/4	10	8	6 7/8	6 5/8
405S	1 7/8	3 1/2	1/2	1/2	2	20	17 3/4	10	8	6 7/8	6 5/8
405U	2 3/8	6 7/8	5/8	5/8	5 1/2	20	17 3/4	10	8	6 7/8	6 5/8
405US	2 1/8	4	1/2	1/2	2 3/4	20	17 3/4	10	8	6 7/8	6 5/8
405T	2 7/8	7	3/4	3/4	5 5/8	20	17 3/4	10	8	6 7/8	6 5/8
405TS	2 1/8	4	1/2	1/2	2 3/4	20	17 3/4	10	8	6 7/8	6 5/8
444	2 3/8	6 7/8	5/8	5/8	5 1/2	22	18 1/2	11	9	7 1/4	7 1/2
444S	2 1/8	4	1/2	1/2	2 3/4	22	18 1/2	11	9	7 1/4	7 1/2
444U	2 7/8	8 3/8	3/4	3/4	7	22	18 1/2	11	9	7 1/4	7 1/2
444US	2 1/8	4	1/2	1/2	2 3/4	22	18 1/2	11	9	7 1/4	7 1/2
444T	3 3/8	8 1/4	7/8	7/8	6 7/8	22	18 1/2	11	9	7 1/4	7 1/2
444TS	2 3/8	4 1/2	5/8	5/8	3	22	18 1/2	11	9	7 1/4	7 1/2
445	2 3/8	6 7/8	5/8	5/8	5 1/2	22	20 1/2	11	9	8 1/4	7 1/2
445S	2 1/8	4	1/2	1/2	2 3/4	22	20 1/2	11	9	8 1/4	7 1/2
445U	2 7/8	8 3/8	3/4	3/4	7	22	20 1/2	11	9	8 1/4	7 1/2
445US	2 1/8	4	1/2	1/2	2 3/4	22	20 1/2	11	9	8 1/4	7 1/2
445T	3 3/8	8 1/4	7/8	7/8	6 7/8	22	20 1/2	11	9	8 1/4	7 1/2
445TS	2 3/8	4 1/2	5/8	5/8	3	22	20 1/2	11	9	8 1/4	7 1/2
504U	2 7/8	8 3/8	3/4	3/4	7 1/4	25	21	12 1/2	10	8	8 1/2
504S	2 1/8	4	1/2	1/2	2 3/4	25	21	12 1/2	10	8	8 1/2
505	2 7/8	8 3/8	3/4	3/4	7 1/4	25	23	12 1/2	10	9	8 1/2
505S	2 1/8	4	1/2	1/2	2 3/4	25	23	12 1/2	10	9	8 1/2

MOTOR FRAME TABLE

Frame No. Series	Third/Fourth Digit of Frame No.							
	D	1	2	3	4	5	6	7
140	3.50	3.00	3.50	4.00	4.50	5.00	5.50	6.25
160	4.00	3.50	4.00	4.50	5.00	5.50	6.25	7.00
180	4.50	4.00	4.50	5.00	5.50	6.25	7.00	8.00
200	5.00	4.50	5.00	5.50	6.50	7.00	8.00	9.00
210	5.25	4.50	5.00	5.50	6.25	7.00	8.00	9.00
220	5.50	5.00	5.50	6.25	6.75	7.50	9.00	10.00
250	6.25	5.50	6.25	7.00	8.25	9.00	10.00	11.00
280	7.00	6.25	7.00	8.00	9.50	10.00	11.00	12.50
320	8.00	7.00	8.00	9.00	10.50	11.00	12.00	14.00
360	9.00	8.00	9.00	10.00	11.25	12.25	14.00	16.00
400	10.00	9.00	10.00	11.00	12.25	13.75	16.00	18.00
440	11.00	10.00	11.00	12.50	14.50	16.50	18.00	20.00
500	12.50	11.00	12.50	14.00	16.00	18.00	20.00	22.00
580	14.50	12.50	14.00	16.00	18.00	20.00	22.00	25.00
680	17.00	16.00	18.00	20.00	22.00	25.00	28.00	32.00

4-70

MOTOR FRAME TABLE (cont.)

Frame No. Series	D	Third/Fourth Digit of Frame No.								
		8	9	10	11	12	13	14	15
140	3.50	7.00	8.00	9.00	10.00	11.00	12.50	14.00	16.00
160	4.00	8.00	9.00	10.00	11.00	12.50	14.00	16.00	18.00
180	4.50	9.00	10.00	11.00	12.50	14.00	16.00	18.00	20.00
200	5.00	10.00	11.00	—	—	—	—	—	—
210	5.25	10.00	11.00	12.50	14.00	16.00	18.00	20.00	22.00
220	5.50	11.00	12.50	—	—	—	—	—	—
250	6.25	12.50	14.00	16.00	18.00	20.00	22.00	25.00	28.00
280	7.00	14.00	16.00	18.00	20.00	22.00	25.00	28.00	32.00
320	8.00	16.00	18.00	20.00	22.00	25.00	28.00	32.00	36.00
360	9.00	18.00	20.00	22.00	25.00	28.00	32.00	36.00	40.00
400	10.00	20.00	22.00	25.00	28.00	32.00	36.00	40.00	45.00
440	11.00	22.00	25.00	28.00	32.00	36.00	40.00	45.00	50.00
500	12.50	25.00	28.00	32.00	36.00	40.00	45.00	50.00	56.00
580	14.50	28.00	32.00	36.00	40.00	45.00	50.00	56.00	63.00
680	17.00	36.00	40.00	45.00	50.00	56.00	63.00	71.00	80.00

MOTOR AND AMPACITY RATINGS

Every motor is considered to be continuous-duty unless the characteristics of the equipment it drives ensure that the motor can't operate under load continuously.

Separate overload protection for motors is to be the motor's nameplate rating.

LOCATIONS

Motors must be installed so that adequate ventilation is provided and so that maintenance operations can be performed.

Open motors with commutators or collector rings must be located so that sparks from the motors can't reach combustible materials.

Enclosed motors must be used in areas where significant amounts of dust are present.

GROUNDING

The frames of portable motors operating at 150 volts or more must be grounded or guarded.

Frames of stationary motors must be grounded (or isolated from grounds) in the following circumstances:

- **When supplied by metal enclosed wiring.**
- **In wet locations, when they are not isolated or guarded.**
- **In hazardous locations.**
- **If any terminal of the motor is over 150 volts to ground.**

All controller enclosures must be grounded, except when attached to portable ungrounded equipment.

Controller-mounted devices must be grounded.

OVERLOAD PROTECTION

Overload protection is not required where it might increase or cause a hazard, as would be the case if overload protection were used on fire pumps.

Continuous-duty motors of more than one horsepower must have overload protection. This protection may be:

OVERLOAD PROTECTION (cont.)

1. An overload device that responds to motor current. These units must be set to trip at 115% of the motor's full-load current (based on the nameplate rating). Motors with a service factor of at least 1.15 or with marked temperature rise of no more than 40°C can have their overloads set to trip at 125% of full-load current.

2. By built-in (at the factory) overload protection.

Motors of one horsepower or less that are nonpermanently installed, manually started, and within sight of their controller are considered to be protected from overload by their branch-circuit protective device. They may be installed on 120-volt circuits of up to 20 amps.

Motors of one horsepower or less that are permanently installed, automatically started, or not within sight of their controllers may be protected from overloads by one of the following methods:

■ **An overload device that responds to motor current.** These units must be set to trip at 115% of the motor's full-load current (based on the nameplate rating). Motors with a service factor of at least 1.15 or with a marked temperature rise of no more than 40° C can have their overloads set to trip at 125% of full-load current.

■ **By overload protection built into the motor.**

Motors that have enough impedance to ensure that overheating is not a threat can be protected by their branch-circuit protection method only.

If normal overload protection is too low to allow a motor to start, it may be increased to 130% of the motor's full-load current.

If fuses are used as overload protection, they must be installed in all ungrounded motor conductors; also in the grounded conductor for cornerground delta systems.

OVERLOAD PROTECTION (cont.)

If trip coils, relays, or thermal cutouts are used as overload protective devices, the requirements for standard motor types are:

- 3-phase AC motors — one overload device must be placed in each phase.
- Single-phase AC or DC, 1 wire grounded — one overload device in the ungrounded conductor.
- Single-phase AC or DC, ungrounded — one overload device in either conductor.
- Single-phase AC or DC, 3 wires, grounded neutral — one overload device in either ungrounded conductor.

Overload protective devices should open enough ungrounded conductors to stop the operation of the motor.

For motors installed on general-purpose branch circuits, overload protection must be as follows:

- Motors of more than one horsepower can be installed on general-purpose branch circuits only when their full-load current is less than 6 amperes, they have overload protection, the branch-circuit protective device marked on any controller is not exceeded, and the overload device is approved for group installation.
- Motors of one horsepower or less can be installed on general-purpose branch circuits without overload protection as long as they comply with the other requirements mentioned above, the full-load current can't exceed 6 amperes, and the branch circuit protective device marked on any controller is not exceeded.

OVERLOAD PROTECTION (cont.)

- When a motor is cord-and-plug-connected, the rating of the plug and receptacle may not be greater than 15 amperes at 125 volts or 10 amperes at 250 volts. If the motor is more than one horsepower, the overload protection must be built into the motor. The branch circuit must be rated per to the rating of the cord and plug.

The branch and overload protections must have enough time delay to allow the motor to start.

Overload protection devices that can restart a motor automatically after tripping are not permitted unless they are approved for use with a specific motor, but *never* allowed if it can cause personal injury.

In situations where the instant shutdown of an overloaded motor would be dangerous to personnel, an alarm should be utilized and an orderly shutdown can be done.

MOTOR CIRCUIT CONDUCTORS

Branch-circuit conductors that supply single motors must have an ampacity of at least 125% of the motor's full-load current rating.

DC motors fed by single-phase rectifiers must have circuit conductors rated at 190% of full-load current for half-wave systems, and 150% of full-load current for full-wave systems.

Conductors connecting secondaries of continuous-duty wound-rotor motors to their controllers must have an ampacity of at least 125% of the full-load secondary current.

MOTOR CONTROLLERS

Suitable controllers are required for all motors.

The branch-circuit protective device can be used as a controller for motors $1/8$ horsepower or less that are normally left running and can't be damaged by overload or starting failure.

Portable motors of $1/3$ horsepower or less may use a plug-and-cord connection as a controller.

Controllers must have horsepower ratings no lower than the horsepower rating of the motor with the following exceptions:

- Controllers that operate motors over 600 volts must have the control circuit voltage visibly marked on the controller.
- Fault-current protection must be provided for each motor operating at over 600 volts.
- All exposed live parts must be protected.

Each motor must have its own controller unless:

A group of motors (600 volts or less) uses a single controller that is rated at no less than the sum of all motors connected to the controller. This applies only in the following scenarios:

1. When a group of motors drives several parts of a single machine.
2. When a group of motors is protected by one overcurrent device.
3. When a group of motors is located in one room, and within sight of the controller.

A controller must be capable of stopping and starting the motor as well as interrupting its lock-rotor current.

The disconnecting means must be located within sight of the controller and the motor with the following exceptions:

1. If the circuit is over 600 volts, the controller disconnecting means can be out of sight from the controller, as long as the controller has a warning

MOTOR CONTROLLERS (cont.)

label that states the location of the disconnecting means is locked in the open position.

2. One disconnecting means can be located next to a group of coordinated controllers on a multi-motor continuous process machine.

The disconnecting means for motors 600 volts or less must be rated at least 115% of the full-load current of the motor.

MOTOR CONTROL CIRCUITS

Motor control circuits that are tapped from the load side of a motor branch-circuit device and control the motor's operation are not considered branch circuits. They can be protected by either a supplementary or a branch-circuit protective device. Control circuits that are not tapped this way are considered signaling circuits.

Motor control conductors must be protected (normally with a fuse in-line) in accordance with Article 430 of the NEC.

When damage to a control circuit would create a hazard, the control circuit must be protected using a raceway, etc., outside of the control enclosure.

When one-side of a motor control circuit is grounded, the circuit must be installed so that accidental grounds won't start the motor.

Motor control circuits must be installed so that they will be shut off from the current when the disconnecting means is in the open position.

CONDUCTORS SUPPLYING MOTORS AND OTHER LOADS

Conductors that supply motors and other loads must have their motor loads computed as specified previously. The other loads must be computed according to their requirements. Then the two loads are added together.

If taps are to be made from feeder conductors, they must terminate into a branch-circuit protective device and must follow these requirements:

- They must have the same ampacity as the feeder conductors *or* be enclosed by a raceway or in a controller.
- They must be no longer than 10 feet *or* have an ampacity of at least one-third of the feeder ampacity.
- They must be protected.
- They must be no longer than 25 feet.

In high-bay manufacturing and production buildings taller than 35 feet from floor to ceiling, taps longer than 25 feet are permitted in these specific situations:

- The tap conductors must be protected from damage and installed in a raceway.
- The tap conductors must terminate in an appropriate circuit breaker or set of fuses.
- The tap conductors must be continuous, with no splices.
- The tap conductors may be run no more than 25 feet horizontally, and no more than 100 feet overall.
- The tap conductors must have an ampacity of at least one-third that of the feeder conductors.
- The tap conductors can't penetrate floors, walls, or ceilings.
- The minimum size of tap conductors is No. 6 AWG copper, or No. 4 AWG aluminum.

Feeders that supply motors in addition to lighting loads must be sized to carry the entire lighting and motor load.

CONDUCTORS SUPPLYING SEVERAL MOTORS

Conductors that supply two or more motors must have an ampacity of no less than the total of the full-load currents of all motors being served plus 25% of the highest-rated motor in the group. If interlock circuitry guarantees that all motors can't be operated at the same time, the calculations can be made based on the largest group of motors that can be operated at anytime.

More than one motor can be connected to the same branch circuit if the following requirements are specifically followed:

- Motors of one horsepower or less can be installed on general purpose branch circuits without overload protection as long as all other requirements are met.
- The full-load current cannot be more than 6 amperes.
- The branch-circuit protective device marked on any controller is not exceeded.

Conductors serving several motors must be provided with a protective device rated no greater than the highest rating of the protective device of any motor in the group *plus* the sum of the full-load currents of the other motors.

SHORT-CIRCUIT AND GROUND-FAULT PROTECTION

Short-circuit and ground-fault devices must be capable of carrying the starting current of the motors they protect.

Normally, protection devices must have a rating of no less than the values given in *Table 430.52* of the NEC. When these values do not correspond with the standard ratings of overcurrent protection devices, the next higher setting can be used.

If the rating given in *Table 430.52* is not sufficient to allow for the motor's starting current, the following methods should be utilized:

- A non-time delay fuse of 600 amps or less can be increased to handle the starting current, but *not* greater than 400% of the motor's full-load current.

SHORT-CIRCUIT AND
GROUND-FAULT PROTECTION (cont.)

- A time delay fuse can be increased to handle the starting current, but *not* greater than 225% of the motor's full-load current.

- The rating of an inverse-time circuit breaker can be increased, but not greater than 400% of full-load currents that are 100 amps or less, or 300% of full-load currents that are over 100 amps.

- The rating of an instantaneous-trip circuit breaker can be increased, but not greater than 1300% of full-load current.

- Fuses rated between 601 and 6000 amps can be increased, but not greater than 300% of the rated full-load current.

Instantaneous-trip circuit breakers can be used as protective devices, only if they are adjustable and part of a listed combination controller that has overload, short-circuit, and ground-fault protection in each conductor.

ADJUSTABLE-SPEED DRIVES

The size of branch circuits or feeders to adjustable-speed drive equipment must be based on the rated current supply to the equipment.

If overload protection is provided by the system controller, no further protection is required.

The disconnecting means for adjustable-speed drive systems may be installed on the supply line, and must be rated at least 115% of the conversion unit's input current.

CHAPTER 5
Maintenance

Preventive maintenance is performed to keep electrical motor and power transmission equipment running with little or no downtime. In the past, the job of a maintenance department was almost always to repair broken equipment and install new equipment.

Because of the high cost of labor, installing new motors is still a top priority of maintenance departments. However, inexpensive monitors are available that can monitor an electrical system for voltage or phase unbalances, voltage losses, phase reversals, over or undervoltages, currents, temperatures, loss of a pump's prime and other conditions that may be signs of major problems.

A preventive maintenance program includes inspection, cleaning, tightening, adjusting and lubricating, keeping equipment dry, and electronically monitoring power circuits. The purpose of a preventive maintenance program is to:

- **Maintain equipment to ensure uninterrupted operations at the highest efficiency for as long as possible.**

- **Protect equipment from dirt, dust, moisture, corrosion, and electrical and mechanical overloads.**

- **Maintain records of all maintenance work to establish future needs and priorities.**

MOTOR MAINTENANCE OPERATIONS

Every Week

1. Examine commutator and brushes.
2. Check oil level in bearings.
3. See that oil rings turn with the shaft.
4. See that the shaft is free of oil and grease from bearings.
5. Examine starter, switch, fuses, and other controls.
6. Start motor and see that it is brought up to speed in normal time.

Every Six Months

1. Clean motor thoroughly, blowing out dirt from windings and wipe commutator and brushes.
2. Inspect commutator clamping ring.
3. Check brushes and renew any that are more than half worn.
4. Examine brush holders and clean them if dirty. Make sure that brushes ride free in the holders.
5. Check brush pressure.
6. Check brush position.
7. Drain, wash out, and renew oil in sleeve bearings.
8. Check grease in ball or roller bearings.
9. Check operating speed or speeds.

10. See that the end play of the shaft is normal.

11. Inspect and tighten connections on the motor and control.

12. Check current input and compare with normal input.

13. Run the motor and examine the drive critically for smooth running, absence of vibration, and worn gears, chains, or belts.

14. Check motor foot bolts, end-shield bolts, pulley, coupling, gear and journal set-screws, and keys.

15. See that all covers, belt and gear guards are in good order, in place, and securely fastened.

Once a Year

1. Clean out and renew grease in ball- or roller-bearing housings.

2. Test insulation by a megohmmeter.

3. Check the air gap.

4. Clean out magnetic dirt that may be hanging on poles.

5. Check clearance between the shaft and journal boxes of sleeve-bearing motors to prevent operation with worn bearings.

6. Clean out undercut slots in the commutator.

7. Examine connections of the commutator and armature coils.

8. Inspect armature bands.

MOTOR REPAIR AND SERVICE RECORD

Motor File #: _____ Serial #: _____
Date Installed: _____ Motor Location: _____

MFR: _____ Type: _____ Frame: _____
HP: _____ Volts: _____ Amps: _____
RPM: _____ Filter Sizes: _____

Date	Operation	Mechanic

5-4

SEMIANNUAL MOTOR MAINTENANCE CHECKLIST

Step	Operation	Mechanic
1	Turn OFF and lock out all power to the motor and its control circuit.	
2	Clean motor exterior and all ventilation ducts.	
3	Check motor's wire raceway.	
4	Check and lubricate bearings as needed.	
5	Check drive mechanism.	
6	Check brushes and commutator.	
7	Check slip rings.	
8	Check motor terminations.	
9	Check capacitors.	
10	Check all mounting bolts.	
11	Check and record line-to-line resistance.	
12	Check and record megohmmeter resistance from L1 to ground.	
13	Check motor controls.	
14	Reconnect motor and control circuit power supplies.	
15	Check line-to-line voltage for balance and level.	
16	Check line current draw against nameplate rating.	
17	Check and record inboard and outboard bearing temperatures.	

ANNUAL MOTOR MAINTENANCE CHECKLIST

Motor File #: _____   Serial #: _____
Date Installed: _____   Motor Location: _____

MFR: _____   Type: _____   Frame: _____
HP: _____   Volts: _____   Amps: _____
RPM: _____   Date Serviced: _____

Step	Operation	Mechanic
1	Turn OFF and lock out all power to the motor and its control circuit.	
2	Clean motor exterior and all ventilation ducts.	
3	Uncouple motor from load and disassemble.	
4	Clean inside of motor.	
5	Check centrifugal switch assemblies.	
6	Check rotors, armatures, and field windings.	
7	Check all peripheral equipment.	
8	Check bearings.	
9	Check brushes and commutator.	
10	Check slip rings.	
11	Reassemble motor and couple to load.	

ANNUAL MOTOR MAINTENANCE CHECKLIST (cont.)

Step	Operation	Mechanic
12	Flush old bearing lubricant and replace.	
13	Check motor's wire raceway.	
14	Check drive mechanism.	
15	Check motor terminations.	
16	Check capacitors.	
17	Check all mounting bolts.	
18	Check and record line-to-line resistance.	
19	Check and record megohmmeter resistance from T1 to ground.	
20	Check and record insulation polarization index.	
21	Check motor controls.	
22	Reconnect motor and control circuit power supplies.	
23	Check line-to-line voltage for balance and level.	
24	Check line current draw against nameplate rating.	
25	Check and record inboard and outboard bearing temperatures.	

LOCATING CIRCUITS

Repairs and upgrades to circuits frequently require the circuit to be identified in the panel that feeds it, even if only to disconnect power. This can be a problem, especially in a poorly-marked, crowded panel. Turning circuits off one-by-one in a panel is not a good option, as some of the circuits may feed sensitive devices.

A clamp-on ammeter and a flashing light can be used to solve this problem quickly and easily. Plug the flashing lamp into any receptacle outlet on the circuit you need to identify, then check the circuits in the panel. The circuit whose current rises and falls along with the flashing of the lamp is the one.

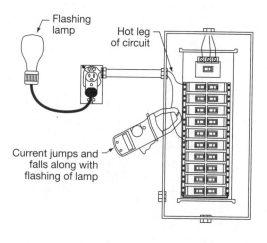

Flashing lamp

Hot leg of circuit

Current jumps and falls along with flashing of lamp

CHECKING CAPACITORS

Capacitors burn out after a limited period of use, and should be checked regularly in applications where burnout would have serious consequences.

The first method in checking a capacitor is to inspect it visually for cracks, bulges or leakage. this provides a fast check for obvious problems, but will not find internal flaws.

A more thorough check is made by using an ohmmeter. (A typical multimeter will have ohmmeter capabilities.) But before this can be done, the capacitor must be discharged. This is done by removing it from the circuit, then connecting a resistor across its leads. A 20 kΩ, 5 watt resistor will usually work, and will discharge the capacitor in 5 – 10 seconds.

Once the capacitor is discharged, the ohmmeter can be connected across the capacitor's terminals. If the capacitor is good, the meter will initially show zero resistance, then slowly move to infinity. If the capacitor is shorted, the meter will show zero, and will not change. If the capacitor is open, the meter will show infinity, and will not change.

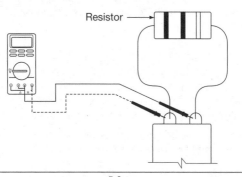

Resistor

UNMARKED 3φ INDUCTION MOTORS

When a motor is in operation for a long period of time, it may cause the markings of the external leads to become defaced. This can also occur to a rebuilt motor that has been in the maintenance shop for awhile. To ensure proper operation, each motor lead must be remarked.

The most common 3φ motors are the single-voltage, three-lead motor and the dual-voltage, nine-lead motor. Both may be internally connected in a wye or delta configuration.

The three leads of a single-voltage, 3φ, three-lead motor can be marked as T1, T2, and T3 in any order. The motor can be connected to the rated voltage and allowed to run. T1 and T3 may be interchanged if the rotation is in the wrong direction.

WYE OR DELTA CONNECTION

A multimeter can be used to determine whether a dual-voltage motor is internally connected in a wye or delta configuration.

A wye-connected motor has three circuits of two leads each (T1-T4, T2-T5, and T3-T6) and one circuit of three leads (T7-T8-T9). A delta-connected motor has three circuits of three leads each (T1-T4-T9, T2-T5-T7, and T3-T6-T8).

Determine the winding circuits (T1-T4, etc.) on an unmarked motor by connecting one meter lead to any motor lead and temporarily connecting the other meter lead to each remaining motor lead. A resistance reading other than infinity indicates a complete circuit.

Caution: Motor must be completely disconnected from circuit before testing for resistance.

WYE-CONNECTED MOTOR

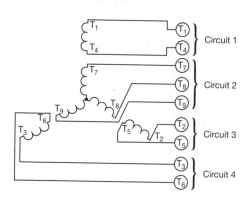

DELTA-CONNECTED MOTOR

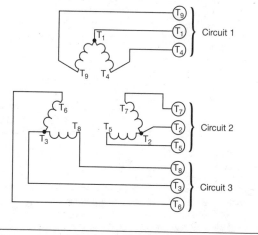

DC MOTOR PERFORMANCE CHARACTERISTICS

Performance Characteristics	Voltage 10% below Rated Voltage		Voltage 10% above Rated Voltage	
	Shunt	Compound	Shunt	Compound
Starting Torque	−15%	−15%	+15%	+15%
Speed	−5%	−6%	+5%	+6%
Current	+12%	+12%	−8%	−8%
Field temperature	Increases	Decreases	Increases	Increases
Armature temperature	Increases	Increases	Decreases	Decreases
Commutator temperature	Increases	Increases	Decreases	Decreases

MAXIMUM ACCELERATION TIME

Frame Number	Maximum Acceleration Time
48 and 56	8 Seconds
143-286	10 Seconds
324-326	12 Seconds
364-505	15 Seconds

The longer a motor takes to accelerate, the higher the temperature rise in the motor. The larger the load, the longer the acceleration time. The maximum recommended acceleration time depends on the motor's frame size. Large motor frames dissipate heat faster than small motor frames.

AC VOLTAGE VARIATION CHARACTERISTICS

Performance Characteristics	10% Above Rated Voltage	10% Below Rated Voltage
Starting current	+10% to +12%	−10% to −12%
Full-load current	−7%	+11%
Motor torque	+20% to +25%	−20% to −25%
Motor efficiency	Little change	Little change
Speed	+1%	−1.5%
Temperature rise	−3°C to −4°C	+6°C to +7°C

Motor performance is affected when the supply voltage varies from a motor's rated voltage. A motor operates satisfactorily with a voltage variation of ±10% from the voltage rating listed on the motor nameplate.

AC FREQUENCY VARIATION CHARACTERISTICS

Performance Characteristics	5% Above Rated Frequency	5% Below Rated Frequency
Starting current	−5% to −6%	+5% to +6%
Full-load current	−1%	+1%
Motor torque	−10%	+11%
Motor efficiency	Slight increase	Slight decrease
Speed	+5%	−5%
Temperature rise	Slight decrease	Slight increase

Motor performance is affected when the frequency varies from a motor's rated frequency. A motor operates satisfactorily with a frequency variation of ±5% from the frequency rating listed on the motor nameplate.

PHASE UNBALANCE AND TEMPERATURE RISE

The loads of three-phase power systems are balanced during installation. An unbalance can begin if additional single-phase loads are added to the system. As these additional loads are energized, one or two of the three-phase lines will begin to carry more or less of the load of the system. This unbalance causes the three-phase lines to move out-of-phase so the lines are no longer 120 electrical degrees apart.

Phase unbalance causes three-phase motors to run at temperatures higher than their listed ratings. The graph below illustrates the relationship between phase unbalance and motor temperature in percentages.

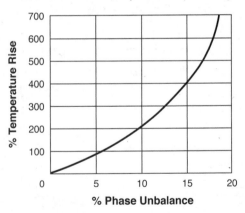

The greater the phase unbalance, the greater the temperature rise. High temperatures produce insulation breakdown and other related problems such as motor inefficiency.

PHASE UNBALANCE DERATING FACTOR

A three-phase motor operating in an unbalanced circuit cannot deliver the horsepower it is rated for. For example, a phase unbalance of 5% will cause a motor to work at 75% of its rated power. This results in the motor having to be derated. The graph below illustrates this relationship.

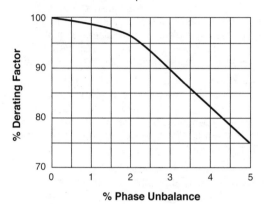

A motor operating on a circuit which has phase unbalance must be derated and will not perform its intended function of work as economically as was originally designed.

It is imperative that the electrical maintenance department check the loads of a three-phase system whenever additions to that system occur.

SINGLE-PHASING CONDITION

Single-phasing is when a three-phase motor is only operating on two phases because one phase has been lost. Single-phasing occurs when one of the lines leading to the motor does not deliver voltage. Single-phasing is the absolute maximum condition of voltage unbalance.

Single-phasing can be distinguished from voltage unbalance by the degree of damage. Voltage unbalance causes less discoloration (blackening), usually over more coils, with little distortion. Single-phasing causes severe distortion and burning to one phase coil.

Severe blackening of one delta winding or two wye windings of the three three-phase windings occurs when a motor has failed due to single phasing. The coil(s) that experienced the voltage loss indicate obvious damage including the insulation on one or possibly two windings.

Possible causes of single-phasing are when one fuse blows, when the switching equipment undergoes a mechanical failure or when lightning blows out one of the lines. Single-phasing can go undetected on most systems because a motor running on two-phase will continue to run in most applications until it burns out.

In most cases, measuring a motor's voltage does not detect a single-phasing condition. The open winding in the motor generates a voltage almost equal to the phase voltage that is lost. When that happens, the open winding acts as the secondary of a transformer, and the two windings connected to power act as the primary. The conditions for single-phasing can be reduced by using correct heater sizes and by using proper size dual-element fuses.

IMPROPER PHASE SEQUENCE (PHASE REVERSAL)

Improper phase sequence is the changing of the sequence of any two phases in a three-phase motor control circuit. Improper phase sequence reverses the motor rotation causing severe mechanical damage and possible injuries to personnel.

Phase reversal can occur when modifications are made to a power distribution system or when maintenance is performed on electrical conductors or switching equipment. Again, care must be taken to properly identify and mark all components of any electrical system undergoing those types of procedures.

In addition, the NEC® requires phase reversal protection on many types of transportation equipment such as escalators, elevators, etc.

VOLTAGE SURGE

A *voltage surge* is a higher-than-normal voltage that temporarily exists on one or more of the power lines. Lightning is a major cause and the surge on a power line can come from a direct hit or induced voltage. Lightning energy can move in both directions on power lines and this traveling surge causes a large voltage rise quickly. The large voltage is impressed on the first couple of turns of the windings, destroying the insulation and causing the motor to burn out. The rest of the windings will appear normal, with very little or no damage. Properly sized (voltage rating) lightning arresters combined with a connection to a good ground will result in maximum voltage surge protection.

For further protection, surge protectors should be installed on the equipment of the system.

VOLTAGE PROBLEMS

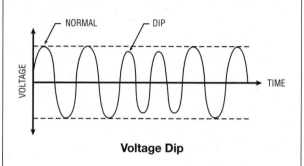

Voltage Dip

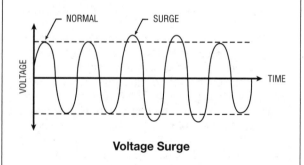

Voltage Surge

VOLTAGE VARIANCE

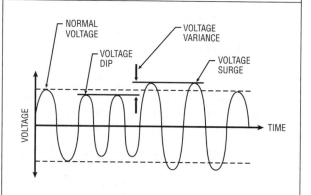

ACCEPTABLE AC LOAD
VOLTAGE RANGES (60HZ)

Rating	Minimum	Maximum
24	22	26
115	109	121
120	114	126
208	197	218
230	218	242
240	228	252
277	263	291
460	436	484
480	456	504

VOLTAGE UNBALANCE

Voltage unbalance occurs when the voltages at different motor terminals are not equal causing one or more motor windings to overheat resulting in thermal deterioration.

A maintenance technician can observe the blackening of one or two of the stator windings which occurs when a motor has failed due to voltage unbalance. The winding with the largest voltage unbalance has the most discoloration.

Voltage unbalance results in a current unbalance. Line voltage should be checked for voltage unbalance periodically or during a service call. Whenever more than 2% voltage unbalance is measured, the following actions should be utilized:

- **The surrounding power system should be examined for excessive loads connected to any one line.**

- **Adjust the excessive load on the motor if the voltage unbalance cannot be corrected.**

- **Change the motor rating by installing an oversized motor if the voltage unbalance cannot be corrected.**

- **Notify your local electric utility of the problem if all else fails.**

FINDING VOLTAGE UNBALANCE

1. Measure the voltage between each incoming power line. Take the readings as follows:

   L1 to L2, L1 to L3 and L2 to L3

2. Add the voltages, then divide by 3 to find the voltage average.

3. Find the voltage deviation by subtracting the voltage average from the voltage with the largest deviation.

4. Apply the following formula to find the voltage unbalance percentage:

$$V_u = \frac{V_d}{V_a} \times 100$$

V_u = voltage unbalance (%)

V_d = voltage deviation (in volts)

V_a = voltage average (in volts)

100 = percentage constant

MOTOR OVERCYCLING

Overcycling is the process of a motor turning on and off repeatedly. Motor starting current is normally 6 to 8 times the full-load running current. The majority of motors are not designed to start more than 10 times per hour. Overcycling begins when a motor is at its operating temperature and still cycles on and off further increasing the temperature and damaging the insulation of the motor.

When your application requires a motor to be cycled often, the following tips may be useful:

- **Utilize a motor with a 50°C rise instead of 40°C.**

- **Utilize a motor with a 1.25 or 1.35 service factor in place of a motor with a factor of 1.00 or 1.15.**

- **Force air over the motor to dissipate the additional heat.**

IMPROPER VENTILATION

Every motor produces heat which must be removed to prevent damage to the insulation. Motors are designed with passages that allow free airflow through the motor removing the heat. Restricting the airflow causes the motor to operate at higher temperatures.

Airflow can be restricted by the accumulation of grease, sludge, grass trimmings, vermin, dirt, etc. Airflow is also restricted if a motor becomes coated with oil from leaks or excessive lubrication. If a motor is placed in an enclosed area it can overheat due to the recirculation of heated air. Adding vents will allow the heated air to dissipate, thus keeping the motor operating at a lower temperature.

EXCESSIVE HEAT

One of the biggest causes of motor problems and failures is excessive heat. Excessive heat will destroy the insulation which, in turn, shorts the windings and renders the motor non-operational.

Insulation life is shortened as the heat in a motor increases beyond the insulation's temperature rating. The temperature rating of motor insulation is listed by classes and is illustrated in the chart below. A motor nameplate almost always lists the insulation class.

MOTOR INSULATION CLASS		
Class	°F	°C
A	221	105
B	266	130
F	311	155
H	356	180

Excessive heat is normally caused by the following conditions:

- **Incorrect motor sizing for the intended application.**
- **Improper ventilation from various sources.**
- **Excessive load on the motor.**
- **Excessive friction from vibration due to improper installation and/or alignment with other mechanical devices coupled to the motor shaft.**
- **Electrical problems such as overloads, phase unbalances, voltage surges or voltage unbalances.**

MOTOR OVERLOADS

The term *overload* refers to a condition that results from applying an excessive load to a motor. Motors try to drive the connected load when power is applied. Obviously, the larger the load, the more power is required to drive the load. Every motor that is manufactured has a rated limit to the load they are able to drive.

For example, a 10 HP, 460 V, 3φ motor should draw no more than 14 amps of current. If a motor begins to draw more than its rating, overloading will begin. Current readings must be taken to determine if an overload problem exists. The chart below illustrates meter readings and their relationship to the motor's listed rating from the nameplate.

Current Rating Of Motor	Meter Readings		
	Fully Loaded Motor	Underloaded Motor	Overloaded Motor
20 A	20 A	12 A	22 A
Nameplate Rating	95 to 105% of Rating	0 to 95% of Rating	105% + of Rating

The even blackening of all of the motor windings occurs when a motor has failed due to overloading. This even discoloration is due to the slow destruction of the windings over a long time period. Normally, no visible damage to the insulation of the motor can be detected.

If a motor is protected properly, overloads should not become a problem. Using correctly-sized heaters in the motor starter will assure that an overload is removed from the motor before any damage occurs.

MEGOHMMETER CONNECTIONS

A megohmmeter (frequently called a *megger*) is used to check insulation strength with a high voltage, usually between 50 and 5000 volts. A megohmmeter may be used to test between circuit conductors or windings, or may test between the conductor or winding to ground.

Since insulation integrity varies over time (and may also vary with other conditions), megger testing is recommended every six months for critical circuits or equipment.

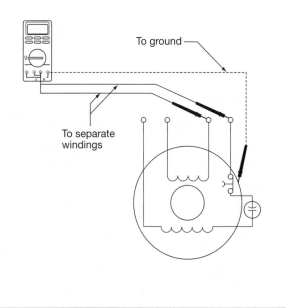

To ground

To separate
windings

OHMMETER CONNECTIONS

An ohmmeter measures the resistance of windings and components in a motor circuit.

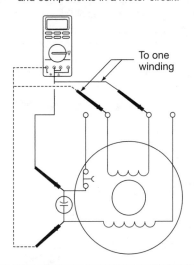

To one winding

Typical Resistance Values at 40°C

Motor Voltage	Minimum Acceptable Resistance
Up to 208 V	100,000 Ω
208-240 V	200,000 Ω
240-600 V	300,000 Ω
600-1000 V	1 MΩ
1000-2400 V	2 MΩ
2400-5000 V	3 MΩ

INSULATION SPOT TESTING

Insulation spot testing verifies insulation quality over the life of the motor. Insulation spot testing should be done every six months using the method below:

A. Measure the resistance of each winding lead to ground. Record the readings after 60 seconds. Then record the lowest meter reading on an insulation spot test graph if all readings are above the minimum acceptable resistance.

B. Connect the motor windings to ground through a 5 kΩ, 5 W resistor. The winding should be connected for 10 times the motor testing time to discharge energy stored in the insulation.

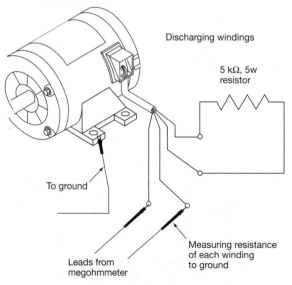

Discharging windings

5 kΩ, 5w resistor

To ground

Measuring resistance of each winding to ground

Leads from megohmmeter

DIELECTRIC ABSORPTION TESTING

A *dielectric absorption test* checks the absorption characteristics of moist or cracked insulation and is performed over a 10-minute period.

A. Using a megohmmeter, measure the resistance of each winding lead to ground. Record the lowest meter reading on a dielectric absorption test graph if all readings are above the minimum acceptable resistance. Record the readings every 10 seconds for the first minute and every minute thereafter for 10 minutes.

B. Discharge the motor windings.

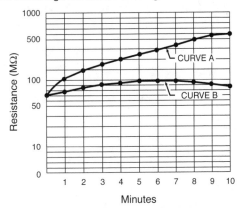

The good insulation is illustrated by Curve A above which shows a gradual increase in resistance levels over time. Cracked or moist insulation (as shown by Curve B) illustrates relatively constant resistance levels over time.

Note: Service the motor if any reading does not meet the minimum acceptable resistance.

POLARIZATION INDEX VALUES

The polarization index indicates the overall condition of motor insulation. Excessive moisture and/or cracks in the insulation are indicated by a low polarization index value.

A polarization index is obtained by dividing the value of the 10-minute reading by the value of the 1-minute reading from a dielectric absorption test.

$$P.I. = \frac{10\text{-minute M}\Omega \text{ Reading}}{1\text{-minute M}\Omega \text{ Reading}}$$

Minimum Acceptable Values	
Insulation	P.I. Value
Class A	1.5
Class B	2.0
Class F	2.0

For example, if the 10-minute reading of Class A insulation is 120 MΩ and the 1-minute reading is 90 MΩ, the polarization index is 1.33 showing the insulation contains excessive moisture or cracks.

INSULATION STEP VOLTAGE TESTING

An *insulation step voltage test* creates electrical stress on internal insulation cracks to reveal damage not found during other insulation tests.

A. Use a megohmmeter setting of 500 Volts to measure the resistance of each winding lead to ground. Observe each reading after one minute and record the lowest reading.

B. Place the meter leads on the winding that has the lowest reading, set the megohmmeter at 1000 Volts, then take a reading starting at 1000 Volts and ending at 5000 Volts, at 500 Volt intervals. Record each reading after one minute has elapsed.

C. Discharge the motor windings.

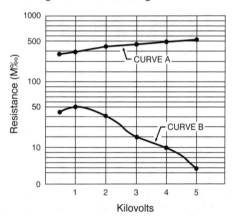

The resistance levels of good insulation, illustrated above by Curve A, remains approximately the same at different voltage levels. The resistance of cracked insulation (Curve B) decreases substantially at different voltage levels.

CHAPTER 6
Troubleshooting

Troubleshooting is a critical skill for anyone who works with motors.

The first step in troubleshooting is to correctly identify the problem—and to do so safely. Some problems will be obvious, but others will not. For the problems that are not obvious, the key to successful troubleshooting (aside from experience) is to be methodical. Start at the power source and follow the current step by step until you can identify the problem. Do not skip around. Follow A to B to C until you find the malfunction.

As mentioned above, safety is key. Never work on motors while wearing anything that could hang onto or into a motor. That includes neck ties, long hair, or anything of the sort. Also, avoid rings and other jewelry which can get caught on an edge or that could accidentally short something out. This is very serious. Even a fairly small motor can have the strength of five or ten horses. That's a lot of power, and it can be very unforgiving.

Be especially careful of DC series motors (or DC compound motors with burnt shunt windings—which is the same thing). They can "run away," and literally blow themselves apart.

And, obviously, you must be careful of the electricity that runs these motors itself.

To summarize the troubleshooting process, always adhere to the following: use common sense, go slowly, carefully and methodically, and think about what you are doing. If there is any complexity to the situation, make drawings of what you are working on as you go.

CONTACTOR AND MOTOR STARTER TROUBLESHOOTING GUIDE

Problem	Possible Cause	Corrective Action
Humming noise	Magnet pole faces misaligned	Realign, Replace magnet assembly if realignment is not possible.
	Too low voltage at coil	Measure voltage at coil. Check voltage rating of coil. Correct any voltage that is 10% less than coil rating.
	Pole face obstructed by foreign object, dirt, or rust	Remove any foreign object and clean as necessary. Never file pole faces.
Loud buzz noise	Shading coil broken	Replace coil assembly.
Controller fails to drop out	Voltage to coil not being removed	Measure voltage at coil. Trace voltage from coil to supply looking for shorted switch or contact if voltage is present.
	Worn or rusted parts causing binding	Clean rusted parts. Replace worn parts.
	Contact poles sticking	Checking for burning or sticky substance on contacts. Replace burned contacts. Clean dirty contacts.
	Mechanical interlock binding	Check to ensure interlocking mechanism is free to move when power is OFF. Replace faulty interlock.
Controller fails to pull in	No coil voltage	Measure voltage at coil terminals. Trace voltage loss from coil to supply voltage if voltage is not present.
	Too low voltage	Measure voltage at coil terminals. Correct voltage level if voltage is less than 10% of rated coil voltage. Check for a voltage drop as large loads are energized.
	Coil open	Measure voltage at coil. Remove coil if voltage is present and correct but coil does not pull in. Measure coil resistance for open circuit. Replace if open.

CONTACTOR AND MOTOR STARTER TROUBLESHOOTING GUIDE (cont.)

Problem	Possible Cause	Corrective Action
Controller fails to pull in (cont.)	Coil shorted	Shorted coil may show signs of burning. The fuse or breakers should trip if coil is shorted. Disconnect one side of coil and reset if tripped. Remove coil and check resistance for short if protection device does not trip. Replace shorted coil. Replace any coil that is burned.
	Mechanical obstruction	Remove any obstructions.
Contacts badly burned or welded	Too high inrush current	Measure inrush current. Check load for problem if higher-than-rated load current. Change to larger controller if load current is correct but excessive for controller.
	Too fast load cycling	Change to larger controller if load cycled ON and OFF repeatedly.
	Too large overcurrent protection device	Size overcurrent protection to load and controller.
	Short circuit	Check fuses or breakers. Clear any short circuit.
	Insufficient contact pressure	Check to ensure contacts are making good connection.
Nuisance tripping	Incorrect overload size	Check size of overload against rated load current. Size up if permissible per NEC®.
	Lack of temperature compensation	Correct setting of overload if controller and load are at different ambient temperatures.
	Loose connections	Check for loose terminal connection.

DIRECT CURRENT MOTOR TROUBLESHOOTING GUIDE

Problem	Possible Cause	Corrective Action
Motor will not start	Blown fuse or open CB	Test the OCPD. If voltage is present at the input, but not the output of the OCPD, the fuse is blown or the CB is open. Check the rating of the OCPD. It should be at least 125% of the motor's FLC.
	Motor overload on starter tripped	Allow overloads to cool. Reset overloads. If reset overloads do not start motor, test the starter.
	No brush contact	Check brushes. Replace, if worn.
	Open control circuit between incoming power and motor	Check for cleanliness, tightness, and breaks. Use a voltmeter to test the circuit starting with the incoming power and moving to the motor terminals. Voltage generally stops at the problem area.
	Excessive load	If the motor is loaded to excess or is jammed, the circuit OCPD will open. Disconnect the load from the motor. If the motor now runs properly, check the load. If the motor does not run and the fuse or CB opens, the problem is with the motor or control circuit. Remove the motor from the control circuit and connect it directly to the power source. If the motor runs properly, the problem is in the control circuit. Check the control circuit. If the motor opens the fuse or CB again, the problem is in the motor. Replace or service the motor.
	Motor shaft does not turn	Disconnect the motor from the load. If the motor shaft still does not turn, the bearings are frozen. Replace or service the motor.
Fuse, CB, or overloads retrip after service	Brush material is too weak or the wrong type for motor's duty rating	Replace with better grade or type of brush. Consult manufacturer if problem continues.
Brushes chip or break	Brush face is overheating and losing brush bonding material	Check for an overload on the motor. Reduce the load as required. Adjust brush holder arms.

DIRECT CURRENT MOTOR TROUBLESHOOTING GUIDE (cont.)

Problem	Possible Cause	Corrective Action
Brushes chip or break	Brush holder is too far from commutator	Too much space between the brush holder and the surface of the commutator allows the brush end to chip or break. Set correct space between brush holder and commutator.
	Brush tension is incorrect	Adjust brush tension so the brush rides freely on the commutator.
Brushes spark	Worn brushes	Replace worn brushes. Service the motor if rapid brush wear, excessive sparking, chipping, breaking, or chattering is present.
	Commutator is concentric	Grind commutator and undercut mica. Replace commutator if necessary.
	Excessive vibration	Balance armature. Check brushes. They should be riding freely.
Rapid brush wear	Wrong brush material, type, or grade	Replace with brushes recommended by manufacturer.
	Incorrect brush tension	Adjust brush tension so the brush rides freely on the commutator.
Motor overheats	Improper ventilation	Clean all ventilation openings. Vacuum or blow dirt out of motor with low-pressure, dry, compressed air.
	Motor is overloaded	Check the load for binding. Check shaft straightness. Measure motor current under operating conditions. If the current is above the listed current rating, remove the motor. Remeasure the current under no-load conditions. If the current is excessive under load but not when unloaded, check the load. If the motor draws excessive current when disconnected, replace or service the motor.

SHADED POLE MOTOR TROUBLESHOOTING GUIDE

Problem	Possible Cause	Corrective Action
Motor will not start	Blown fuse or open CB	Test OCPD. If voltage is present at the input, but not the output of the OCPD, the fuse is blown or the CB is open. Check the rating of the OCPD. It should be at least 125% of the motor's FLC.
	Motor overload on starter tripped	Allow overloads to cool. Reset overloads. If reset overloads do not start the motor, test the starter.
	Low or no voltage applied to motor	Check the voltage at the motor terminals. The voltage must be present and within 10% of the motor nameplate voltage. If voltage is present at the motor but the motor is not operating, remove the motor from the load the motor is driving. Reapply power to the motor. If the motor runs, the problem is with the load. If the motor does not run, the problem is with the motor. Replace or service the motor.
	Open control circuit between incoming power and motor	Check for cleanliness, tightness, and breaks. Use a voltmeter to test the circuit starting with the incoming power and moving to the motor terminals. Voltage generally stops at the problem area.
Fuse, CB, or overloads retrip after service	Excessive load	If the motor is loaded to excess or jammed, the circuit OCPD will open. Disconnect the load from the motor. If the motor now runs properly, check the load. If the motor does not run and the fuse or CB opens, the problem is with the motor or control circuit. Remove the motor from the control circuit and connect it directly to the power source. If the motor runs properly, the problem is in the control circuit. Check the control circuit. If the motor opens the fuse or CB again, the problem is in the motor. Replace or service the motor.
Excessive noise	Unbalanced motor or load	An unbalanced motor or load causes vibration, which causes noise. Realign the motor and load. Check for excessive end play or loose parts. If the shaft is bent, replace the rotor or motor.
	Dry or worn bearings	Dry or worn bearings cause noise. Bearings may be dry due to dirty oil, oil not reaching the shaft, or motor overheating. Oil bearings as recommended. If noise remains, replace the bearings or motor.
	Excessive grease	Ball bearings that have excessive grease may cause bearings to overheat. Overheated bearings cause noise. Remove excess grease.

SPLIT-PHASE MOTOR TROUBLESHOOTING GUIDE

Problem	Possible Cause	Corrective Action
Motor will not start	Thermal cutout switch is open	Reset the thermal switch. Caution: Resetting the thermal switch may automatically start the motor.
	Blown fuse or open CB	Test the OCPD. If voltage is present at the input, but not the output of the OCPD, the fuse is blown or the CB is open. Check the rating of the OCPD. It should be at least 125% of the motor's FLC.
	Motor overload on starter tripped	Allow overloads to cool. Reset overloads. If reset overloads do not start the motor, test the starter.
	Low or no voltage applied to motor	Check the voltage at the motor terminals. The voltage must be present and within 10% of the motor nameplate voltage. If voltage is present at the motor but the motor is not operating, remove the motor from the load the motor is driving. Reapply power to the motor. If the motor runs, the problem is with the load. If the motor does not run, the problem is with the motor. Replace or service the motor.
	Open control circuit between incoming power and motor	Check for cleanliness, tightness, and breaks. Use a voltmeter to test the circuit starting with the incoming power and moving to the motor terminals. Voltage generally stops at the problem area.
	Starting winding not receiving power	Check the centrifugal switch to make sure it connects the starting winding when the motor is OFF.
Fuse, CB, or overloads retrip after service	Blown fuse or open CB	Test the OCPD. If voltage is present at the input, but not the output of the OCPD, the fuse is blown or the CB is open. Check the rating of the OCPD. It should be at least 125% of the motor's FLC.
	Motor overload on starter tripped	Allow overloads to cool. Reset overloads. If reset overloads do not start the motor, test the starter.

SPLIT-PHASE MOTOR TROUBLESHOOTING GUIDE (cont.)

Problem	Possible Cause	Corrective Action
Fuse, CB, or overloads retrip after service	Low or no voltage applied to motor	Check the voltage at the motor terminals. The voltage must be present and within 10% of the motor nameplate voltage. If voltage is present at the motor but the motor is not operating, remove the motor from the load the motor is driving. Reapply power to the motor. If the motor runs, the problem is with the load. If the motor does not run, the problem is with the motor. Replace or service the motor.
	Open control circuit between incoming power and motor	Check for cleanliness, tightness, and breaks. Use a voltmeter to test the circuit starting with the incoming power and moving to the motor terminals. Voltage generally stops at the problem area.
	Motor shaft does not turn	Disconnect the motor from the load. If the motor shaft still does not turn, the bearings are frozen. Replace or service the motor.
Motor produces electric shock	Broken or disconnected ground strap	Connect or replace ground strap. Test for proper ground.
	Hot power lead at motor connecting terminals is touching motor frame	Disconnect the motor. Open the motor terminal box and check for poor connections, damaged insulation, or leads touching the frame. Service and test motor for ground.
	Motor winding shorted to frame	Remove, service, and test motor.
Motor overheats	Starting windings are not being removed from circuit as motor accelerates	When the motor is turned OFF, a distinct click should be heard as the centrifugal switch closes.
	Improper ventilation	Clean all ventilation openings. Vacuum or blow dirt out of motor with low-pressure, dry, compressed air.

SPLIT-PHASE MOTOR TROUBLESHOOTING GUIDE (*cont.*)

Problem	Possible Cause	Corrective Action
Motor overheats	Motor is overloaded	Check the load for binding. Check shaft straightness. Measure motor current under operating conditions. If current is above the listed current rating, remove the motor. Remeasure the current under no-load conditions. If the current is excessive under load but not when unloaded, check the load. If the motor draws excessive current when disconnected, replace or service the motor.
	Dry or worn bearings	Dry or worn bearings cause noise. The bearings may be dry due to dirty oil, oil not reaching the shaft, or motor overheating. Oil the bearings as recommended. If noise remains, replace the bearings or the motor.
	Dirty bearings	Clean or replace bearings.
Excessive noise	Excessive end-play	Check and play by trying to move the motor shaft in and out. Add end-play washers as required.
	Unbalanced motor or load	An unbalanced motor or load causes vibration, which causes noise. Realign the motor and load. Check for excessive end play or loose parts. If the shaft is bent, replace the rotor or motor.
	Dry or worn bearings	Dry or worn bearings cause noise. The bearings may be dry due to dirty oil, oil not reaching the shaft, or motor overheating. Oil the bearings as recommended. If noise remains, replace the bearings or the motor.
	Excessive grease	Ball bearings that have excessive grease may cause the bearings to overheat. Overheated bearings cause noise. Remove any excess grease.

THREE-PHASE MOTOR TROUBLESHOOTING GUIDE

Problem	Possible Cause	Corrective Action
Motor will not start	Wrong motor connections	Most 3φ motors are dual-voltage. Check for proper motor connections.
	Blown fuse or open CB	Test the OCPD. If voltage is present at the input, but not the output of the OCPD, the fuse is blown or the CB is open. Check the rating of the OCPD. It should be at least 125% of the motor's FLC.
	Motor overload on starter tripped	Allow overloads to cool. Reset overloads. If reset overloads do not start the motor, test the starter.
	Low or no voltage applied to motor	Check the voltage at the motor terminals. The voltage must be present and within 10% of the motor nameplate voltage. If voltage is present at the motor but the motor is not operating, remove the motor from the load the motor is driving. Reapply power to the motor. If the motor runs, the problem is with the load. If the motor does not run, the problem is with the motor. Replace or service the motor.
	Open control circuit between incoming power and motor	Check for cleanliness, tightness, and breaks. Use a voltmeter to test the circuit starting with the incoming power and moving to the motor terminals. Voltage generally stops at the problem area.
Fuse, CB, or overloads retrip after service	Power not applied to all three lines	Measure voltage at each power line. Correct any power supply problems.
	Blown fuse or open CB	Test the OCPD. If voltage is present at the input, but not the output of the OCPD, the fuse is blown or the CB is open. Check the rating of the OCPD. It should be at least 125% of the motor's FLC.
	Motor overload on starter tripped	Allow overloads to cool. Reset overloads. If reset overloads do not start the motor, test the starter.

THREE-PHASE MOTOR TROUBLESHOOTING GUIDE (cont.)

Problem	Possible Cause	Corrective Action
Fuse, CB, or overloads retrip after service	Low or no voltage applied to motor	Check the voltage at the motor terminals. The voltage must be present and within 10% of the motor nameplate voltage. If voltage is present at the motor but the motor is not operating, remove the motor from the load the motor is driving. Reapply power to the motor. If the motor runs, the problem is with the load. If the motor does not run, the problem is with the motor. Replace or service the motor.
	Open control circuit between incoming power and motor	Check for cleanliness, tightness, and breaks. Use a voltmeter to test the circuit starting with the incoming power and moving to the motor terminals. Voltage generally stops at the problem area.
	Motor shaft does not turn	Disconnect the motor from the load. If the motor shaft still does not turn, the bearings are frozen. Replace or service the motor.
Motor overheats	Motor is single phasing	Check each of the 3∅ power lines for correct voltage.
	Improper ventilation	Clean all ventilation openings. Vacuum or blow dirt out of motor with low-pressure, dry, compressed air.
	Motor is overloaded	Check the load for binding. Check shaft straightness. Measure motor current under operating conditions. If the current is above the listed current rating, remove the motor. Remeasure the current under no-load conditions. If the current is excessive under load but not when unloaded, check the load. If the motor draws excessive current when disconnected, replace or service the motor.

FAULTY SOLENOID PROBLEMS

Problem	Possible Causes	Comments
Failure to operate when energized	Complete loss of power to solenoid	Normally caused by blown fuse or control circuit problem.
	Low voltage applied to the solenoid	Voltage should be at least 85% of solenoid's rated value.
	Burned out solenoid coil	Normally evident by pungent odor caused by burnt insulation.
	Shorted coil	Normally a fuse is blown and continues to blow when changed.
	Obstruction of plunger movement	Normally caused by a broken part, misalignment, or the presence of a foreign object.
	Excessive pressure on solenoid plunger	Normally caused by excessive system pressure in solenoid-operated valves.
Failure to operate spring-return solenoids when de-energized	Faulty control circuit	Normally a problem of the control circuit not disengaging the solenoid's hold or memory circuit.
	Obstruction of plunger movement	Normally caused by a broken part, misalignment, or the presence of a foreign object.
	Excessive pressure on solenoid plunger	Normally caused by excessive system pressure in solenoid-operated valves
Failure to operate electrically-operated return solenoids when de-energized	Complete loss of power to solenoid	Normally caused by a blown fuse or control circuit problem.
	Low voltage applied to solenoid	Voltage should be at least 85% of solenoid's rated value.

FAULTY SOLENOID PROBLEMS (cont.)

Problem	Possible Causes	Comments
Failure to operate electrically operated return solenoids when de-energized	Burned out solenoid coil	Normally evident by pungent odor caused by burnt insulation.
	Obstruction of plunger movement	Normally caused by broken part, misalignment, or presence of a foreign object
	Excessive pressure on solenoid plunger	Normally caused by excessive system pressure in solenoid-operated valves
Noisy operation	Solenoid housing vibrates	Normally caused by loose mounting screws.
	Plunger pole pieces do not make flush contact	An air gap may be present causing the plunger to vibrate. These symptoms are normally caused by foreign matter.
Erratic operation	Low voltage applied to the solenoid	Voltage should be at least 85% of the solenoid's rated voltage.
	System pressure may be low or excessive	Solenoid size is inadequate for the application.
	Control circuit is not operating properly	Conditions on the solenoid have increased to the point where the solenoid cannot deliver the required force.

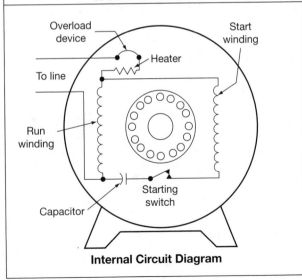

MECHANICAL START

Split-phase motor
with bad capacitor

To line

Normal
Rotational
direction

String

SPLIT-PHASE MOTOR

Overload
device

Heater

Start
winding

To line

Run
winding

Starting
switch

Capacitor

Internal Circuit Diagram

6-14

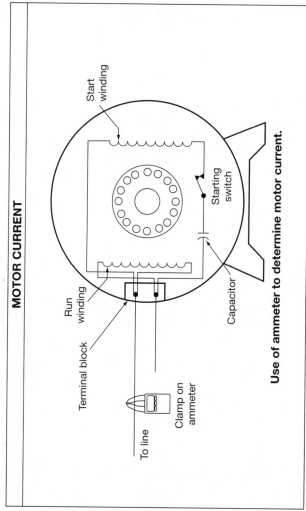

Use of ammeter to determine motor current.

DEFECTIVE POLE

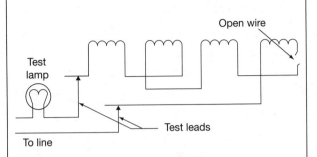

**Finding a defective pole. If the circuit is open,
the lamp will not light.**

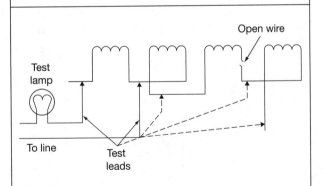

Determining which pole is open-circuited.

START WINDING

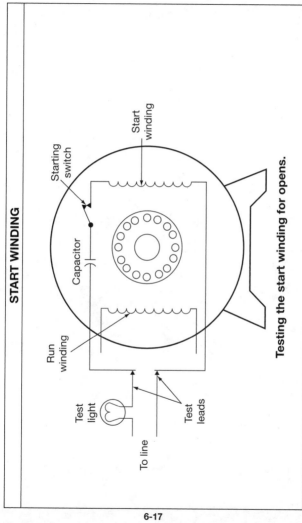

Testing the start winding for opens.

STATER

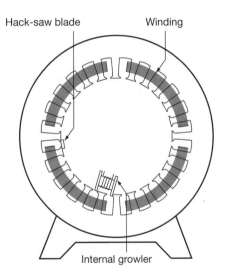

Hack-saw blade   Winding

Internal growler

Growler method of testing for shorts in the stator.

GROUNDED WINDING

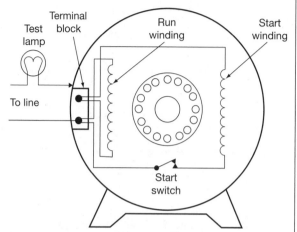

Test lamp

Terminal block

Run winding

Start winding

To line

Start switch

Connect one test lead to the winding and the other test lead to the core. A lighted lamp will indicate a ground.

DISASSEMBLY

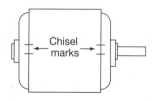

Chisel marks

Mark both the motor end plates and the frame before disassembling.

SHORTED COIL

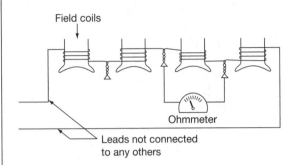

Field coils

Ohmmeter

Leads not connected
to any others

Using an ohmmeter to detect a shorted coil.

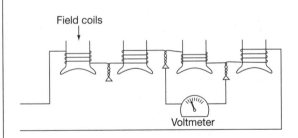

Field coils

Voltmeter

Using a voltmeter to locate a shorted coil.

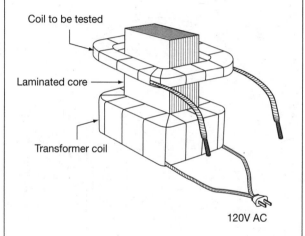

Coil to be tested

Laminated core

Transformer coil

120V AC

Using a transformer to detect a shorted coil.

OPENS

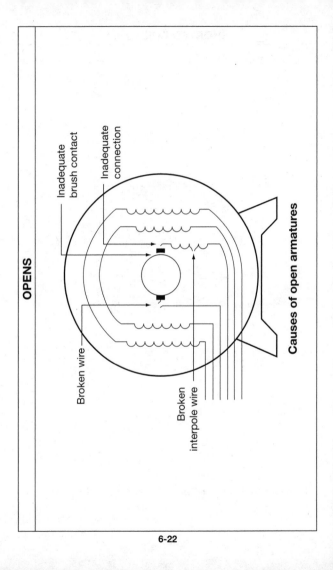

Inadequate brush contact

Inadequate connection

Broken wire

Broken interpole wire

Causes of open armatures

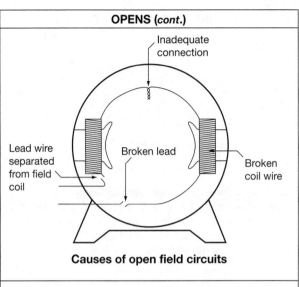

Inadequate
connection

Lead wire
separated
from field
coil

Broken lead

Broken
coil wire

Causes of open field circuits

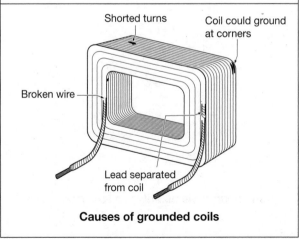

Shorted turns

Coil could ground
at corners

Broken wire

Lead separated
from coil

Causes of grounded coils

INTERPOLE POLARITY

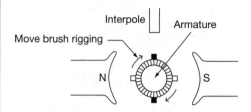

Two-pole motor

**All connections are removed except the
armature and interpole. The brushes are shifted
90 degrees. If the armature turns in the same
direction in which the brushes were moved,
the polarity is correct.**

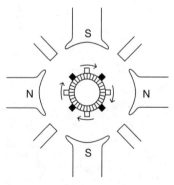

Correct interpole polarity in a four-pole motor.

BRUSHES

The sandpaper must be pulled
only in the direction of the rotation

sandpaper

Wrong

Right

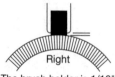

Right

The brush holder is 1/16"
maximum from commutator.

The brush tip will break

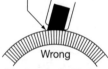

Wrong

The brush stud is too
far from the commutator.

Brush tip will break and burn.

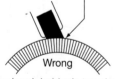

Wrong

The brush holder is too close
to the commutator.

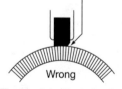

Wrong

The brush holder is too far
from commutator.

Positioning carbon brushes.

COMPOUND MOTORS

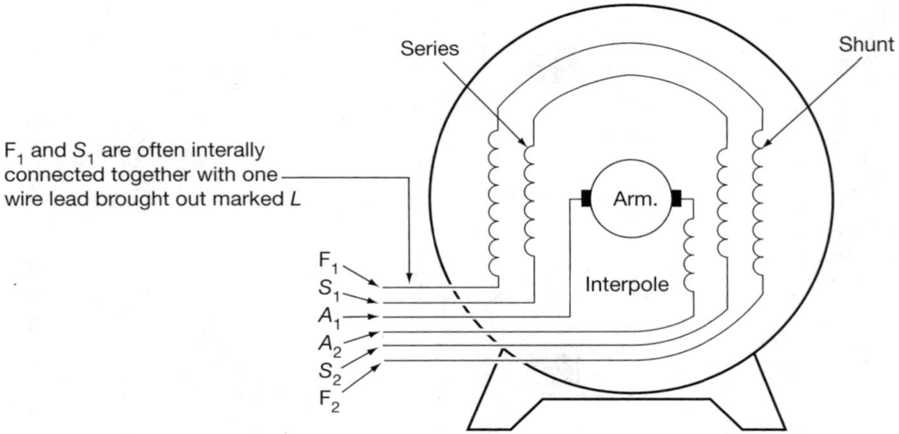

F$_1$ and S$_1$ are often interally connected together with one wire lead brought out marked L

Shunt

Series

Arm.

Interpole

F$_1$
S$_1$
A$_1$
A$_2$
S$_2$
F$_2$

The lead markings of a typical compound motor.

6-26

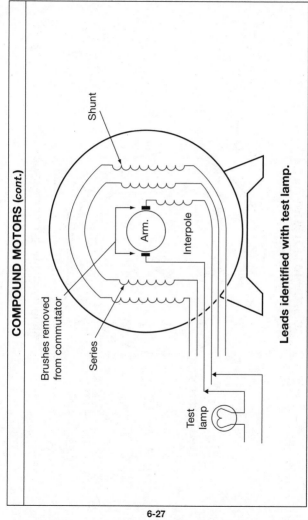

COMPOUND MOTORS (cont.)

Shunt

Brushes removed
from commutator

Series

Arm.

Interpole

Test
lamp

Leads identified with test lamp.

6-27

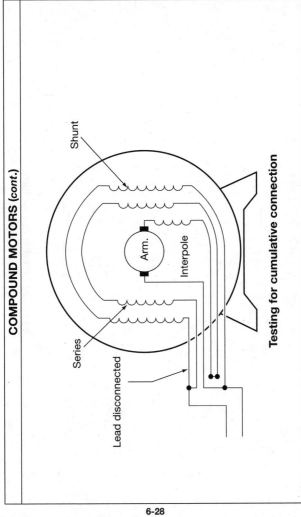

Shunt

Arm.

Interpole

Series

Lead disconnected

Testing for cumulative connection

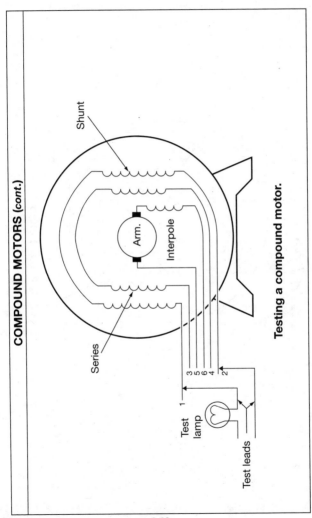

Testing a compound motor.

DC OPENS

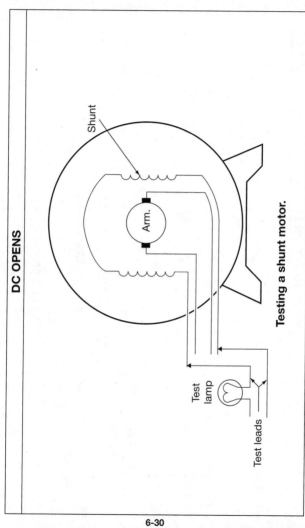

Testing a shunt motor.

Shunt

Arm.

Test lamp

Test leads

6-30

DC OPENS *(cont.)*

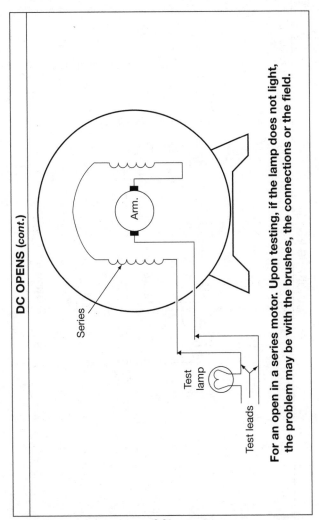

Series

Arm.

Test lamp

Test leads

For an open in a series motor. Upon testing, if the lamp does not light, the problem may be with the brushes, the connections or the field.

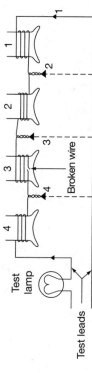

Test lamp

Test leads

Broken wire

Testing field coils

DC GROUNDS

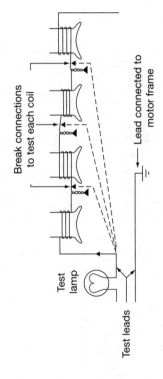

Break connections to test each coil

Lead connected to motor frame

Test lamp

Test leads

Each coil ground tested

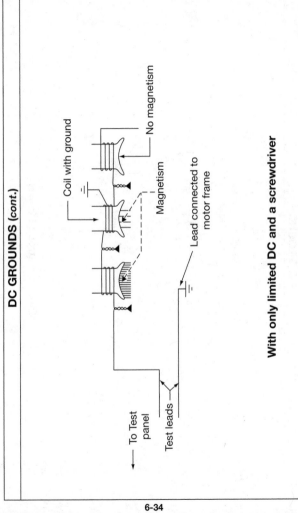

Coil with ground

No magnetism

Magnetism

Lead connected to motor frame

To Test panel

Test leads

With only limited DC and a screwdriver

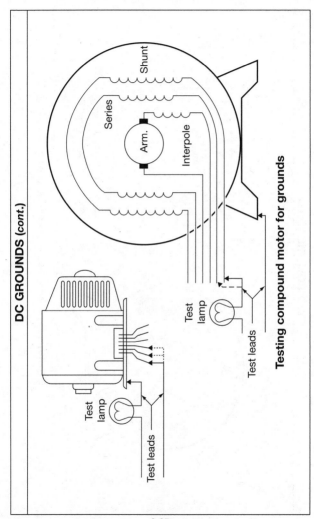

Shunt

Series

Arm.

Interpole

Test
lamp

Test leads

Testing compound motor for grounds

Test
lamp

Test leads

6-35

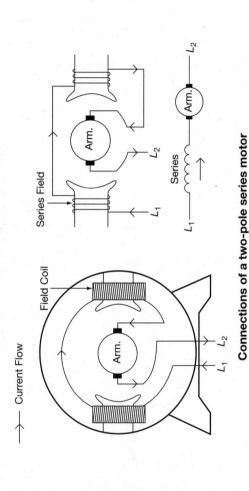

TWO-POLE MOTORS

Field Coil

Arm.

L_2
L_1

→ Current Flow

Connections of a two-pole series motor

Series Field

Arm.

L_1

L_2

Arm.

Series

L_2

L_1

TWO-POLE MOTORS (cont.)

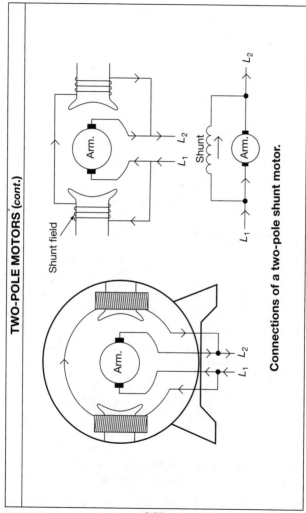

Shunt field

Arm.

L_1 L_2

Shunt

Arm.

L_1 L_2

Connections of a two-pole shunt motor.

POLARITY

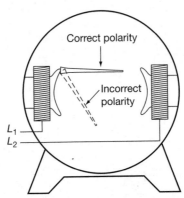

Testing the polarity of a field coil with a nail.

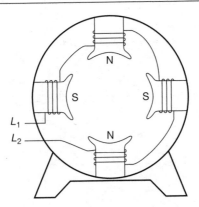

North and south poles alternate
in a four-pole motor.

POLARITY (*cont.*)

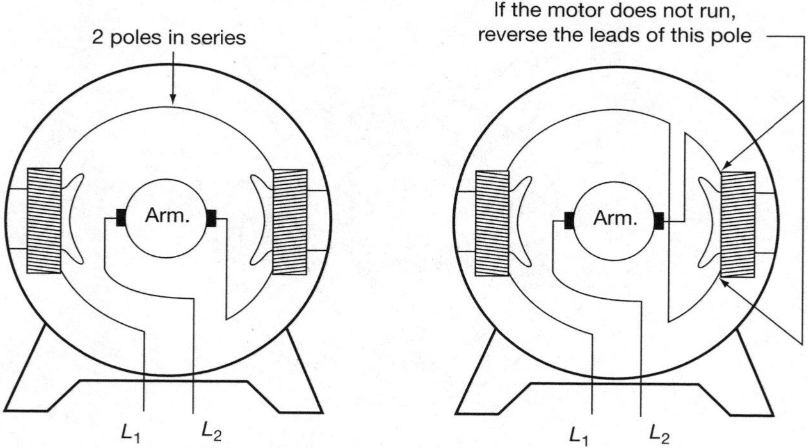
2 poles in series
Arm.
L_1
L_2
If the motor does not run, reverse the leads of this pole
Arm.
L_1
L_2

Test for correct field polarity on a small two-pole motor.

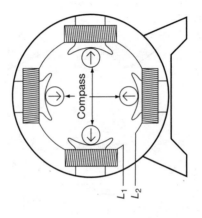

Compass

L_1
L_2

In a four-pole motor, adjacent poles must have opposite polarity.

AC GROUNDS

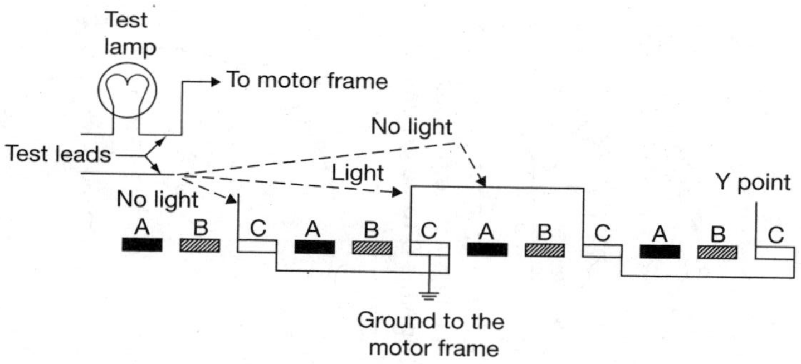

Test
lamp
To motor frame
Test leads
No light
Light
No light
A
B
C
A
B
C
A
B
C
A
B
C
Y point
Ground to the
motor frame

Locating grounded group with test light.

AC GROUNDS (cont.)

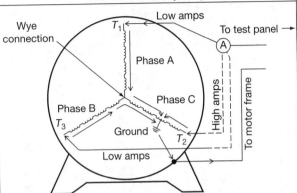

Testing wye motor to locate the grounded phase.
T_2 has the highest amp reading, showing
phase C to be grounded.

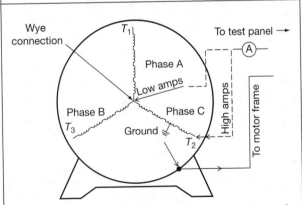

Testing phase C of a wye motor to locate the end
closer to the grounded coil.

AC GROUNDS (cont.)

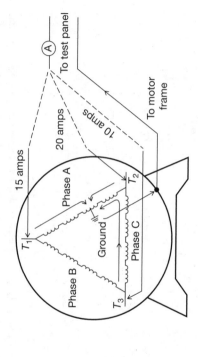

Testing series-delta motor for a grounded phase. T_2 has the highest amp reading, and T_1 is second highest showing the grounding to be in phase A close to T_2.

AC GROUNDS (cont.)

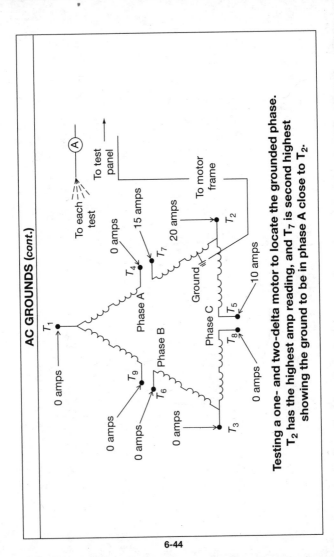

Testing a one- and two-delta motor to locate the grounded phase. T_2 has the highest amp reading, and T_7 is second highest showing the ground to be in phase A close to T_2.

AC OPENS

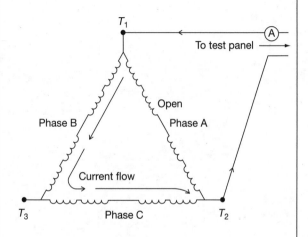

Open phase in a delta connection. Open phase will have the lower amp reading.

AC OPENS (cont.)

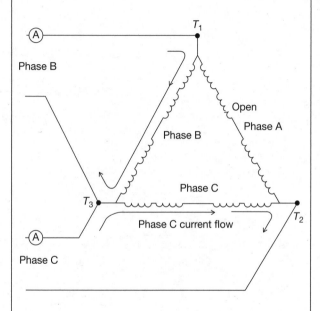

More current will flow when testing across the good phases than across the open phase.

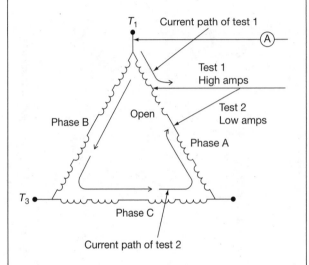

T_1

Current path of test 1

Ⓐ

Test 1
High amps

Test 2
Low amps

Phase B

Open

Phase A

T_3

Phase C

Current path of test 2

**Current in test 1 is high because it goes through
only one group. Current in test 2 goes through
most of the groups and is low.**

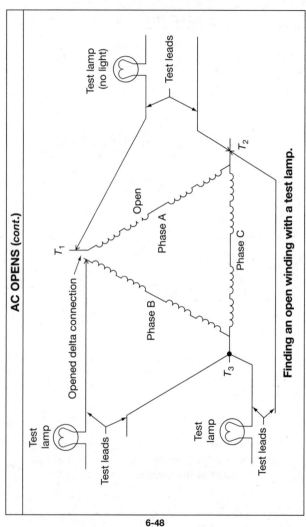

AC OPENS (cont.)

Test lamp (no light)

Test leads

T_1
T_2

Open
Phase A
Phase C

Opened delta connection
Phase B

T_3

Test lamp
Test leads

Test lamp
Test leads

Finding an open winding with a test lamp.

6-48

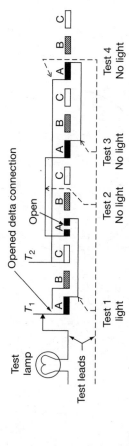

An open group with a test lamp on a delta connection.

Wye
point

Test
lamp

Test leads

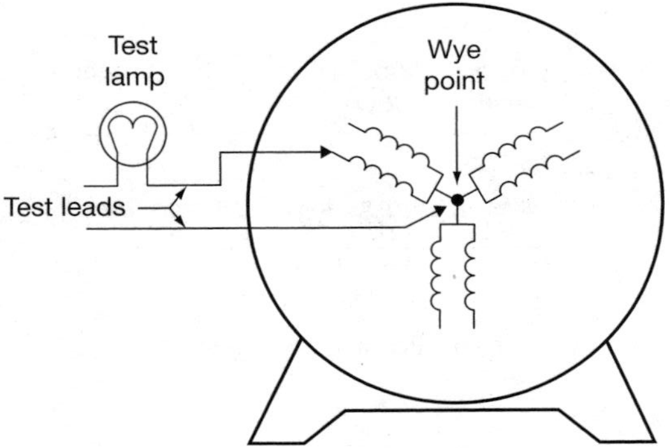

Locating an open in a two-parallel wye motor using a test lamp.

AC OPENS (*cont.*)

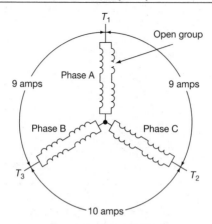

The open phase will draw less current.

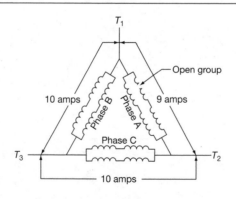

The open phase will draw less current.

SHORTS

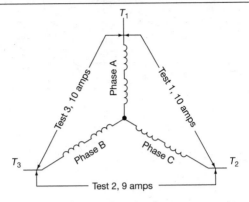

The readings indicate that
phase A may have a short.

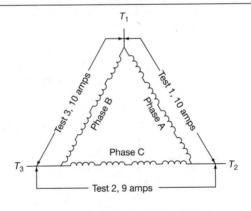

The readings indicate that phase A
may have a short.

FUSES

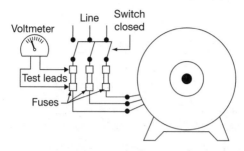

If a fuse is blown, there will be a voltage reading.

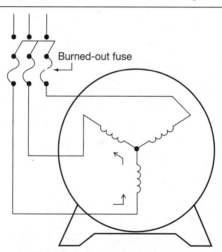

Star-connected motor with burned-out fuse in one phase. Higher current through the other two phases will burn out the coils.

FUSES (cont.)

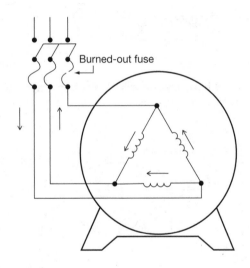

Burned-out fuse

**Delta-connected motor with burned-out fuse
in one phase. High current will flow in
one of the other phases.**

MECHANICAL

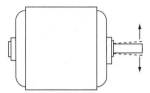

Shaft movement indicates a worn bearing.

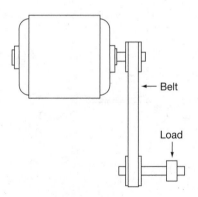

Belt

Load

**Disconnect belt and try to move load to see
if it is free to turn.**

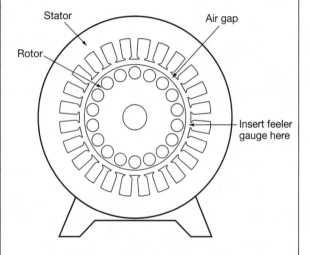

Stator

Rotor

Air gap

Insert feeler gauge here

**Air gap should be the same around the entire motor.
This is checked with a feeler gauge.**

CHAPTER 7
Power Transmission

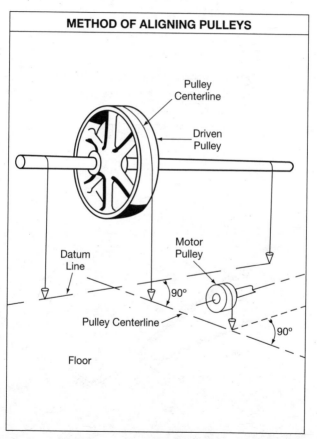

METHOD OF ALIGNING PULLEYS

Pulley Centerline

Driven Pulley

Motor Pulley

Datum Line

90°

Pulley Centerline

90°

Floor

CALCULATING PULLEY DIAMETER

A pulley can be used to change the speed of a motor-driven device. To calculate the driven machine pulley diameter use the formula:

$$PD_m = \frac{PD_d \times N_d}{N_m}$$

PD_m = driven machine pulley diameter (in.)

PD_d = drive pulley diameter (in.)

N_d = motor drive speed (rpm)

N_m = driven machine speed (rpm)

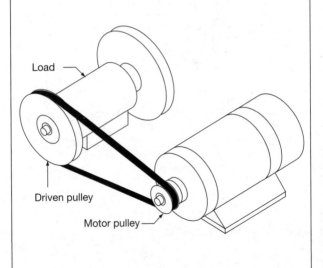

Load

Driven pulley

Motor pulley

PULLEY AND GEAR CALCULATIONS

For single reduction or increase of speed by means of belting where the speed at which each shaft should run is known, and one pulley is in place:

Multiply the diameter of the pulley which you have by the number of revolutions per minute that its shaft makes; divide this product by the speed in revolutions per minute at which the second shaft should run. The result is the diameter of pulley to use.

Where both shafts with pulleys are in operation and the speed of one is known:

Multiply the speed of the shaft by diameter of its pulley and divide this product by diameter of pulley on the other shaft. The result is the speed of the second shaft.

Where a countershaft is used, to obtain size of main driving or driven pulley, or speed of main driving or driven shaft, it is necessary to calculate, as above, between the known end of the transmission and the countershaft, then repeat this calculation between the countershaft and the unknown end.

A set of gears of the same pitch transmits speeds in proportion to the number of teeth they contain. Count the number of teeth in the gear wheel and use this quantity instead of the diameter of pulley, mentioned above, to obtain number of teeth cut in unknown gear, or speed of second shaft.

FORMULAS FOR FINDING PULLEY SIZES

$$d = \frac{D \times S}{S'} \qquad D = \frac{d \times S'}{S}$$

d = diameter of driven pulley.

D = diameter of driving pulley.

S = number of revolutions per minute of driving pulley.

S' = number of revolutions per minute of driven pulley.

FORMULAS FOR FINDING GEAR SIZES

$$n = \frac{N \times S}{S'} \qquad N = \frac{n \times S'}{S}$$

n = number of teeth in pinion (driving gear).

N = number of teeth in gear (driven gear).

S = number of revolutions per minute of pinion.

S' = number of revolutions per minute of gear.

FORMULA TO DETERMINE SHAFT DIAMETER

The formula for determining the size of steel shaft for transmitting a given power at a given speed is as follows:

$$\text{diameter of shaft in inches} = \sqrt{\frac{K \times HP}{RPM}}$$

when HP = the horsepower to be transmitted

RPM = speed of shaft

K = factor which varies from 50 to 125 depending on type of shaft and distance between supporting bearings.

For line shaft having bearings 8 feet apart:

K = 90 for turned shafting

K = 70 for cold-rolled shafting

FORMULA TO DETERMINE BELT LENGTH

The following formula is used to determine the length of belting:

$$\text{length of belt} = \frac{3.14\,(D + d)}{2} + 2\sqrt{X^2 + \left(\frac{D - d}{2}\right)^2}$$

when D = diameter of large pulley

d = diameter of small pulley

X = distance between centers of shafting

GEAR REDUCER FORMULAS

Output Torque

$$O_T = I_T \times R_R \times R_E$$

where

O_T = output torque (in lb.-ft.)

I_T = input torque (in lb.-ft.)

R_R = gear reducer ratio

R_E = reducer efficiency %

Output Speed

$$O_S = \frac{I_S}{R_R} \times R_E$$

where

O_s = output speed (in rpm)

I_S = input speed (in rpm)

R_R = gear reducer ratio

R_E = reducer efficiency %

Output Horsepower

$$O_{HP} = I_{HP} \times R_E$$

where

O_{HP} = output horsepower

I_{HP} = input horsepower

R_E = reducer efficiency %

MOTOR TORQUE FORMULAS

Torque

$$T = \frac{HP \times 5252}{rpm}$$

where

T = torque

HP = horsepower

5252 = constant $\left(\frac{33,000 \text{ lb-ft}}{\pi \times 2} = 5252\right)$

rpm = revolutions per minute

Starting Torque

$$T = \frac{HP \times 5252 \times \%}{rpm}$$

where

HP = horsepower

5252 = constant $\left(\frac{33,000 \text{ lb-ft}}{\pi \times 2} = 5252\right)$

rpm = revolutions per minute

$\%$ = motor class percentage

Nominal Torque Rating

$$T = \frac{HP \times 63,000}{rpm}$$

where

T = nominal torque rating (in lb-in)

$63,000$ = constant

HP = horsepower

rpm = revolutions per minute

ADJUSTING BELT TENSION

A belt must be tight enough not to slip, but not so tight that it overloads the motor bearings. Belt tension is checked by placing a straightedge from pulley to pulley and measuring the amount of deflection at the midpoint. As a general rule, belt deflection should equal $\frac{1}{64}$" per inch of span. Belt tension is adjusted by moving the drive motor away from or closer to the driven device.

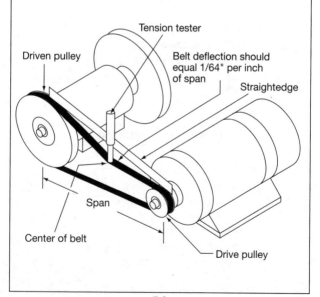

Tension tester

Driven pulley

Belt deflection should equal 1/64" per inch of span

Straightedge

Span

Center of belt

Drive pulley

V-BELTS

No. 0 Section
"2L"

No. 1 Section
"3L"

No. 2 Section
"4L"
A

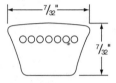

No. 3 Section
"5L"
B

V-BELT/MOTOR SIZE

3/8"
9.5 mm

Up to .76 kW
1 HP

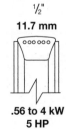

1/2"
11.7 mm

.56 to 4 kW
5 HP

21/32"
16.7 mm

2.5 kW and up
3 HP

STANDARD "V" BELT LENGTHS

A BELTS		
Standard Belt No.	Pitch Length	Outside Length
A26	27.3	28.0
A31	32.3	33.0
A35	36.3	37.0
A38	39.3	40.0
A42	43.3	44.0
A46	47.3	48.0
A51	52.3	53.0
A55	56.3	57.0
A60	61.3	62.0
A68	69.3	70.0
A75	76.3	77.0
A80	81.3	82.0
A85	86.3	87.0
A90	91.3	92.0
A96	97.3	98.0
A105	106.3	107.0
A112	113.3	114.0
A120	121.3	122.0
A128	129.3	130.0

B BELTS		
Standard Belt No.	Pitch Length	Outside Length
B35	36.8	38.0
B38	39.8	41.0
B42	43.8	45.0
B46	47.8	49.0
B51	52.8	54.0
B55	56.8	58.0
B60	61.8	63.0
B68	69.8	71.0
B75	76.8	78.0
B81	82.8	84.0
B85	86.8	88.0
B90	91.8	93.0
B97	98.8	100.0
B105	106.8	108.0
B112	113.8	115.0
B120	121.8	123.0
B128	129.8	131.0
B136	137.8	139.0
B144	145.8	147.0
B158	159.8	161.0
B173	174.8	176.0
B180	181.8	183.0
B195	196.8	198.0
B210	211.8	213.0
B240	240.3	241.5
B270	270.3	271.5
B300	300.3	301.5

C BELTS		
Standard Belt No.	Pitch Length	Outside Length
C51	53.9	55.0
C60	62.9	64.0
C68	70.9	81.0
C75	77.9	79.0
C81	83.9	85.0
C85	87.9	89.0
C90	92.9	94.0
C96	98.9	100.0
C105	107.9	109.0
C112	114.9	116.0
C120	122.9	124.0
C128	130.9	132.0
C136	138.9	140.0
C144	146.9	148.0
C158	160.9	162.0
C162	164.9	166.0
C173	175.9	177.0
C180	182.9	184.0
C195	197.9	199.0
C210	212.9	214.0
C240	240.9	242.0
C270	270.9	272.0
C300	300.9	302.0
C360	360.9	362.0
C390	390.9	392.0
C420	420.9	422.0

D BELTS		
Standard Belt No.	Pitch Length	Outside Length
D120	123.3	125.0
D128	131.3	133.0
D144	147.3	149.0
D158	1613	163.0
D162	165.3	167.0
D173	176.3	178.0
D180	183.3	185.0
D195	198.3	200.0
D210	213.3	215.0
D240	240.8	242.0
D270	270.8	272.5
D300	300.8	302.5
D330	330.8	332.5
D360	360.8	362.5
D390	390.8	392.5
D420	420.8	422.5
D480	480.8	482.5
D540	540.8	542.5
D600	600.8	602.5

E BELTS			E BELTS		
Standard Belt No.	Pitch Length	Outside Length	Standard Belt No.	Pitch Length	Outside Length
E180	184.5	187.5	E360	361.0	364.0
E195	199.5	202.5	E390	391.0	394.0
E210	214.5	217.5	E420	421.0	424.0
E240	241.0	244.0	E480	481.0	484.0
E270	271.0	274.0	E540	541.0	544.0
E300	301.0	304.0	E600	601.0	604.0
E330	331.0	334.0			

STANDARD "V" BELT LENGTHS (cont.)

3V Belts		5V Belts		8V Belts	
3V250	25.0	5V500	50.0	8V1000	100.0
3V265	26.5	5V530	53.0	8V1060	106.0
3V280	28.0	5V560	56.0	8V1120	112.0
3V300	30.0	5V600	60.0	8V1180	118.0
3V315	31.5	5V630	63.0	8V1250	125.0
3V335	33.5	5V670	67.0	8V1320	132.0
3V355	35.5	5V710	71.0	8V1400	140.0
3V375	37.5	5V750	75.0	8V1500	150.0
3V400	40.0	5V800	80.0	8V1600	160.0
3V425	42.5	5V850	85.0	8V1700	170.0
3V450	45.0	5V900	90.0	8V1800	180.0
3V475	47.5	5V950	95.0	8V1900	190.0
3V500	50.0	5V1000	100.0	8V2000	200.0
3V530	53.0	5V1060	106.0	8V2120	212.0
3V560	56.0	5V1120	112.0	8V2240	224.0
3V600	60.0	5V1180	118.0	8V2360	236.0
3V630	63.0	5V1250	125.0	8V2500	250.0
3V670	67.0	5V1320	132.0	8V2650	265.0
3V710	71.0	5V1400	140.0	8V2800	280.0
3V750	75.0	5V1500	150.0	8V3000	300.0
3V800	80.0	5V1600	160.0	8V3150	315.0
3V850	85.0	5V1700	170.0	8V3350	335.0
3V900	90.0	5V1800	180.0	8V3550	355.0
3V950	95.0	5V1900	190.0	8V3750	375.0
3V1000	100.0	5V2000	200.0	8V4000	400.0
3V1060	106.0	5V2120	212.0	8V4250	425.0
3V1120	112.0	5V2240	224.0	8V4500	450.0
3V1180	118.0	5V2360	236.0	8V5000	500.0
3V1250	128.0	5V2500	250.0		
3V1320	132.0	5V2650	265.0		
3V1400	140.0	5V2800	280.0		
		5V3000	300.0		
		5V3150	315.0		
		5V3350	335.0		
		5V3550	355.0		

If the 60-inch "B" section belt shown is made 3/10 of an inch longer, it will be code marked 53 rather than 50. If made 3/10 shorter, it will be marked 47. While both have the belt number B60 they cannot be used in a set because of the difference in length.

TYPICAL CODE MARKING

B60   MANUFACTURER'S NAME   50

NOMINAL
SIZE AND LENGTH

LENGTH
CODE NUMBER

HORSEPOWER CAPACITIES OF
LIGHT 4-PLY NYLON-STITCHED BELTS

Belt Speed	Pulley Diameter (inches)			
(Feet Per Minute)	1"	2"	3"	6"
1000	0.45	0.64	0.76	0.83
2000	0.94	1.27	1.38	1.56
3000	1.28	1.81	2.07	2.35
5000	1.89	2.73	3.18	3.75
6000	1.92	2.91	3.35	4.05
7000	1.93	2.94	3.57	4.51
8000	1.87	2.98	3.63	4.72
9000		2.92	3.68	4.79
10000			3.64	4.85
12000				4.81

HORSEPOWER CAPACITIES OF
MEDIUM 4-PLY NYLON-STITCHED BELTS

Belt Speed	Pulley Diameter (inches)			
(Feet Per Minute)	1"	2"	3"	6"
1000	0.60	0.90	1.03	1.25
2000	1.15	1.67	1.95	2.33
3000	1.59	2.44	2.85	3.50
5000	1.80	3.09	3.80	4.85
6000	1.86	3.30	4.17	5.31
7000	1.82	3.35	4.25	5.54
8000		3.33	4.27	5.67
9000			4.21	5.71
10000				5.62

HORSEPOWER CAPACITIES OF HEAVY 4-PLY NYLON-STITCHED BELTS

Belt Speed	Pulley Diameter (inches)			
(Feet Per Minute)	1"	2"	3"	6"
1000	0.90	1.40	1.70	2.21
2000	1.66	2.06	3.21	4.19
3000	2.35	3.82	4.75	6.15
5000	3.03	5.51	7.00	9.49
6000	2.99	6.17	8.21	11.30
7000		6.10	8.39	12.20
8000			8.31	12.48
9000				12.25

HORSEPOWER CAPACITIES OF MEDIUM 4-PLY WOVEN ENDLESS COTTON BELTS

Belt Speed	Pulley Diameter (inches)			
(Feet Per Minute)	1-1½"	3"	4"	6"
500	0.25	0.35	0.40	0.50
1000	0.50	0.70	0.80	1.00
1500	0.80	1.00	1.30	1.50
2000	1.00	1.40	1.60	2.00
2500	1.30	1.80	2.00	2.30
3000	1.50	2.00	2.30	2.60
3500	1.80	2.40	2.50	3.00
4000	2.00	2.80	2.80	3.50
4500	2.10	3.20	3.20	4.10
5000	2.20	3.30	3.30	4.50

HORSEPOWER CAPACITIES PER INCH OF WIDTH OF REGULAR SINGLE-PLY DACRON BELTS

Belt Speed (Feet Per Minute)	Pulley Diameter (inches)						
	½"	1"	1½"	2"	3"	4"	6"
1000	0.5	1.0	1.0	1.2	1.4	1.5	1.6
2000	1.5	2.0	2.1	2.2	2.5	2.7	2.9
3000	1.9	2.4	3.0	3.3	3.9	4.1	4.4
4000	2.3	2.9	3.7	4.2	4.8	5.2	5.7
5000	2.5	3.2	4.2	4.7	5.5	6.0	6.7
6000	2.6	3.5	4.6	5.1	6.0	6.7	7.5
7000	2.6	3.6	4.8	5.4	6.5	7.4	8.4
8000	2.5	3.7	5.0	5.6	6.8	7.8	8.9
9000	2.3	3.8	5.1	5.7	6.9	8.0	9.2
10000	2.0	3.8	5.2	5.8	7.0	8.2	9.4
12000		3.6	5.2	5.8	7.0	8.2	9.5
15000		3.3	5.0	5.7	6.9	8.2	9.5
18000		3.0	4.8	5.6	6.9	8.1	9.5

HORSEPOWER CAPACITIES PER INCH OF WIDTH OF MEDIUM SINGLE-PLY DACRON BELTS

Belt Speed (Feet Per Minute)	Pulley Diameter (inches)						
	½"	1"	1½"	2"	3"	4"	6"
1000	0.33	0.528	0.66	0.79	0.924	0.99	1.056
2000	0.99	1.32	1.38	1.45	1.65	1.78	1.914
3000	1.25	1.58	1.98	2.17	2.594	2.70	2.90
4000	1.51	1.91	2.44	2.77	3.168	3.43	3.762
5000	1.65	2.11	2.77	3.10	3.63	3.96	4.42
6000	1.71	2.31	3.03	3.36	3.96	4.42	4.95
7000	1.71	2.37	3.16	3.56	4.29	4.884	5.54
8000	1.65	2.44	3.30	3.69	4.488	5.148	5.874
9000	1.51	2.50	3.36	3.76	4.554	5.28	6.072
10000	1.32	2.50	3.43	3.82	4.62	5.412	6.138
12000		2.37	3.43	3.82	4.62	5.412	6.20
15000		2.17	3.30	3.76	4.554	5.412	6.20
18000		1.98	3.16	3.69	4.554	5.346	6.27

HORSEPOWER CAPACITIES PER INCH OF WIDTH OF LIGHT SINGLE-PLY DACRON BELTS

Belt Speed (Feet Per Minute)	Pulley Diameter (inches)						
	½"	1"	1½"	2"	3"	4"	6"
1000	0.20	0.32	0.40	0.48	0.560	0.600	0.640
2000	0.60	0.80	0.84	0.88	1.00	1.08	1.16
3000	0.76	0.96	1.20	1.32	1.56	1.64	1.76
4000	0.92	1.16	1.48	1.68	1.92	2.08	2.28
5000	1.00	1.28	1.68	1.88	2.20	2.40	2.68
6000	1.04	1.40	1.84	2.04	2.40	2.68	3.00
7000	1.04	1.44	1.92	2.16	2.60	2.96	3.36
8000	1.00	1.48	2.00	2.24	2.72	3.12	3.56
9000	0.92	1.52	2.04	2.28	2.76	3.20	3.68
10000	0.80	1.52	2.08	2.32	2.80	3.28	3.72
12000		1.44	2.08	2.32	2.80	3.28	3.76
15000		1.32	2.00	2.28	2.76	3.28	3.76
18000		1.20	1.92	2.24	2.76	3.24	3.80

HORSEPOWER CAPACITIES OF ⅜" DIAMETER WOUND ENDLESS ROUND BELTS

Belt Speed (Feet Per Minute)	Pitch Diameter of Pulley (inches)						
	3"	3.4"	3.8"	4.2"	4.6"	5.0"
400	0.46	0.54	0.61	0.66	0.70	0.74
800	0.76	1.10	1.27	1.40	1.51	1.61
1200	1.02	1.27	1.46	1.62	1.76	1.87
1600	1.22	1.56	1.82	2.03	2.21	2.36
2000	1.30	1.80	2.13	2.40	2.62	2.80
2400	1.51	2.01	2.41	2.70	3.00	3.20
2800	1.59	2.20	2.61	3.00	3.30	3.55
3200	1.61	2.30	2.80	3.23	3.60	3.87
3600	1.60	2.34	3.00	3.54	3.98	4.13
4000	1.50	2.34	3.00	3.54	3.98	4.35
4800	1.16	2.16	2.95	3.60	4.03	4.57
5200	0.90	2.00	2.80	3.50	4.10	4.60
5600	0.54	1.71	2.64	3.38	4.00	4.50
6000	0.12	1.40	2.40	3.20	3.80	4.40

CORRECTION FACTORS FOR SMALL PULLEY ANGLES OF CONTACT LESS THAN 180°										
Belt Type	45°	90°	110°	120°	130°	140°	150°	160°	170°
Nylon Stitched & Woven Endless			0.70	0.74	0.78	0.83	0.87	0.91	0.96
Single-Ply Dacron	0.30	0.60	0.70	0.74	0.78	0.83	0.87	0.91	0.96
Endless Round Belts		0.69	0.79	0.82	0.86	0.89	0.92	0.95	0.97

CORRECTION FACTORS FOR BELT VARIATIONS

Belt Type	Light 4-ply	Heavy 4-ply	Light 6-ply	Medium 6-ply	Heavy 6-ply	Medium 8-ply	Heavy 8-ply
Nylon Stitched			1.12	1.17		1.28	1.40
Woven Endless Cotton	0.33	1.50	0.50	1.50			

HORSEPOWER CAPACITIES OF ¼" DIAMETER BRAIDED ENDLESS ROUND BELTS

Belt Speed (Feet Per Minute)	Pulley Diameter, U-grooves (inches)				
	1½"	2"	3"	4"	5"
400	0.17	0.22	0.23	0.32	0.35
800	0.21	0.36	0.37	0.65	0.69
1200	0.33	0.45	0.45	0.79	0.83
1600	0.41	0.54	0.55	0.83	0.86
2000	0.44	0.58	0.59	0.98	1.04
2400	0.52	0.69	0.63	1.12	1.16
2800	0.54	0.72	0.68	1.24	1.28
3200	0.58	0.77	0.72	1.44	1.49
3600	0.63	0.84	0.74	1.59	1.60
4000	0.67	0.90	0.68	1.59	1.68
4800	0.69	0.92	0.59	1.57	1.70
5200	0.58	0.90	0.45	1.57	1.71
5600	0.63	0.85	0.28	1.56	1.70
6000	0.60	0.80	0.28	1.47	1.70

HORSEPOWER CAPACITIES OF ⅜" DIAMETER BRAIDED ENDLESS ROUND BELTS

Belt Speed (Feet Per Minute)	Pulley Diameter, U-grooves (inches)			
	2"	3"	4"	5"
500	0.45	0.50	0.70	0.78
750	0.72	0.80	1.44	1.61
1000	0.90	1.00	1.75	1.83
1250	1.08	1.20	1.85	1.90
1500	1.20	1.30	2.20	2.30
2000	1.25	1.40	2.50	2.58
2500	1.35	1.52	2.75	2.85
3000	1.45	1.61	3.20	3.30
3500	1.47	1.65	3.54	3.65
4000	1.37	1.52	3.54	3.72
4500	1.18	1.30	3.50	3.75
5000	0.91	1.00	3.50	3.77
5500	0.54	0.60	3.40	3.80
6000	0.27	0.30	3.30	3.75

HORSEPOWER CAPACITIES OF 9/16" DIAMETER WOUND ENDLESS ROUND BELTS

Belt Speed (Feet Per Minute)	Pitch Diameter of Pulley (inches)					
	5.0"	5.4"	5.8"	6.2"	6.6"	7.0"
400	0.94	1.00	1.10	1.20	1.22	1.25
800	1.60	1.80	1.92	2.10	2.17	2.25
1200	2.20	2.45	2.70	2.90	3.00	3.20
1600	2.70	3.00	3.30	3.70	3.80	4.00
2000	3.10	3.50	3.90	4.40	4.50	4.70
2400	3.50	4.00	4.40	4.95	5.10	5.40
2800	3.75	4.30	4.80	5.50	5.70	6.00
3200	3.95	4.61	5.20	5.90	6.10	6.50
3600	4.10	4.80	5.40	6.00	6.50	6.90
4000	4.10	4.90	5.60	6.20	6.80	7.20
4400	3.96	4.45	5.70	6.34	6.90	7.50
4800	3.75	4.30	5.60	6.40	7.00	7.60
5200	3.40	4.00	5.40	6.20	6.90	7.60
5600	2.96	3.60	5.10	6.00	6.70	7.40
6000	2.40	3.00	4.70	5.60	6.40	7.10

METHOD OF CONNECTION: ADJUSTABLE-SPEED DRIVE

Motor	Drive	Space required	Speed Variation	Relative cost
Constant Speed	Variable pitch motor pulley	Same as V-belt	1¼:1	Low
Constant Speed	Separate unit with variable-diameter pulleys	Considerable	3½:1	Medium
Constant Speed	Variable-diameter pulley unit built on motor housing	Moderate	3½:1	Medium
Constant Speed	Variable-position rollers	Small	10:1	High
Adjustable Speed	Direct coupling	Small	4:1	High

METHOD OF CONNECTION: CONSTANT-SPEED DRIVE

Drive	Relative Noise	Space Required	Speed Ratio	Peak Load	Cost Comparison
Direct coupling, standard motor	Zero	Small; motor must extend out from end of drive shaft	Zero	Limited only by motor	1 (considered as 100 percent)
Direct coupling, general-purpose gear motor	Slight	Small; angle drive obtainable in small motors	Any	Full load	2 times standard motor
Direct coupling, special-purpose gear motor	Slight	Small; angle drive obtainable in small motors	Any	150 percent of full load	2½ times standard motor
Direct coupling heavy-duty gear motor	Slight	Small; angle drive obtainable in small motors	Any	200 percent of full load	3 times standard motor

Drive	Relative Noise	Space Required	Speed Ratio	Peak Load	Cost Comparison
V belt	Zero	Moderate	10:1 in fractional-horsepower sizes; down to 4:1 for 50 hp and up	Any by multiplying motor horsepower by a service factor	1½ times standard motor
Flat belt	Considerable	Considerable	Same as V belt	Same as V belt	Slightly higher than for standard motor
Chain	Considerable	Moderate	8:1	Same as V belt	1½ times standard motor
Gear	High	Small	10:1, single reduction; 100:1, double reduction	Same as V belt	1½ times standard motor

MAXIMUM HORSEPOWER: TWO-BEARING MOTORS WITH CHAIN DRIVES

Motor speed, RPM	Maximum HP	Motor speed, RPM	Maximum HP	Motor speed, RPM	Maximum HP
1,700-1,800	5	1,150-1,200	25	720-750	75
1,440-1,500	10	850-900	50		

HORSEPOWER LIMITS: TWO-BEARING MOTORS WITH BELT DRIVE

Motor speed, RPM	Flat-belt drive, maximum HP	V-belt drive, maximum HP	Motor speed, RPM	Flat-belt drive, maximum HP	V-belt drive, maximum HP
1,700-1,800	40	60; ball-bearing, 75	680-720	200	300
1,440-1,500	40	60; ball-bearing, 75	560-600	200	300
1,150-1,200	75	100; ball-bearing, 125	500-514	150	300
850-900	125	200	440-450	150	250

SPEED LIMITATIONS: BELT, GEAR, AND CHAIN DRIVES

Under normal operating conditions for motors not provided with outboard bearings

Full-Load RPM of motor		Max HP rating of motor
Above	Including	
Flat-belt drive		
2,400	3,600	20
1,800	2,400	30
1,200	1,800	40
900	1,200	75
750	900	125
720	750	150
560	720	200
V-Belt Drive		
2,400	3,600	20
1,800	2,400	40
1,200	1,800	75
900	1,200	125
750	900	200
720	750	250
590	720	300
Gear Drive		
1,500	1,800	7½
1,200	1,500	15
900	1,200	25
750	900	50
560	750	75
Chain Drive		
2,400	3,600	20
1,800	2,400	40
1,200	1,800	75
900	1,200	125
750	900	200
720	750	250
560	720	300

NOTES: Figures are based on the use of pulleys and will be less than those given when motors are belted to low-speed drives such as countershafts.
Outboard bearings are recommended for belted motors in frame sizes of 250 hp, 575 to 600 rpm and larger.

About The Author

Paul Rosenberg has an extensive background in the construction, data, electrical, HVAC and plumbing trades. He is a leading voice in the electrical industry with years of experience from an apprentice to a project manager. Paul has written for all of the leading electrical and low voltage industry magazines and has authored more than 30 books.

In addition, he wrote the first standard for the installation of optical cables (ANSI-NEIS-301) and was awarded a patent for a power transmission module. Paul currently serves as contributing editor for *Power Outlet Magazine*, teaches for Iowa State University and works as a consultant and expert witness in legal cases. He speaks occasionally at industry events.